# TSA Past Paper Worked Solutions

# Volume Two

Copyright © 2019 *UniAdmissions*. All rights reserved.

ISBN 978-1-912557-29-5

No part of this publication may be reproduced or transmitted in any form or by any means, electronic or mechanical, including photocopying, recording, or by any information retrieval system without prior written permission of the publisher. This publication may not be used in conjunction with or to support any commercial undertaking without the prior written permission of the publisher.

Published by *RAR Medical Services Limited*
www.uniadmissions.co.uk
info@uniadmissions.co.uk
Tel: 0208 068 0438

TSA is a registered trademark of Cambridge Assessment, which was not involved in the production of, and does not endorse, this book. The authors and publisher are not affiliated with TSA. The answers and explanations given in this book are purely the opinions of the authors rather than an official set of answers.

The information offered in this book is purely advisory and any advice given should be taken within this context. As such, the publishers and authors accept no liability whatsoever for the outcome of any applicant's TSA performance, the outcome of any university applications or for any other loss. Although every precaution has been taken in the preparation of this book, the publisher and author assume no responsibility for errors or omissions of any kind. Neither is any liability assumed for damages resulting from the use of information contained herein. This does not affect your statutory rights.

# TSA Past Paper Worked Solutions

Joseph Nelson

Rohan Agarwal

# About the Author

**Rohan** is the **Director of Operations** at *UniAdmissions* and is responsible for its technical and commercial arms. He graduated from Gonville and Caius College, Cambridge and is a fully qualified doctor. Over the last five years, he has tutored hundreds of successful Oxbridge and Medical applicants. He has also authored ten books on admissions tests and interviews.

Rohan has taught physiology to undergraduates and interviewed medical school applicants for Cambridge. He has published research on bone physiology and writes education articles for the Independent and Huffington Post. In his spare time, Rohan enjoys playing the piano and table tennis.

## The Basics .................................................................................. 6

## 2013 ............................................................................................ 10
### Section 1 ................................................................................ 10
### Section 2 ................................................................................ 30

## 2014 ............................................................................................ 38
### Section 1 ................................................................................ 38
### Section 2 ................................................................................ 63

## 2015 ............................................................................................ 75
### Section 1 ................................................................................ 75
### Section 2 .............................................................................. 104

## 2016 .......................................................................................... 115
### Section 1 .............................................................................. 115
### Question 2: C ....................................................................... 115
### Question 3: C ....................................................................... 115
### Question 4: E ....................................................................... 116
### Question 5 :B ....................................................................... 116
### Section 2 .............................................................................. 135

## 2017 .......................................................................................... 143
### Section 1 .............................................................................. 143
### Section 2 .............................................................................. 158

## 2018 .......................................................................................... 165
### Section 1 .............................................................................. 165
### Section 2 .............................................................................. 176

## 2019 .......................................................................................... 180
### Section 1 .............................................................................. 180
### Section 2 .............................................................................. 198

# The Basics

## What are TSA Past Papers?

Thousands of students take the TSA exam each year. These exam papers are then released online to help future students prepare for the exam. Before 2013, these papers were not publically available meaning that students had to rely on the specimen papers and other resources for practice. However, since their release in 2013, TSA past papers have become an invaluable resource in any student's preparation.

## Where can I get TSA Past Papers?

**This book does not include TSA past paper questions** because it would be over 1,000 pages long! However, TSA past papers from 2008 are available for free from the official TSA website. To save you the hassle of downloading lots of files, we've put them all into one easy-to-access (and free!) folder for you at **www.uniadmissions.co.uk/tsa-past-papers**.

At the time of publication, the 2017 paper has not been released so this book only contains answers for 2008 – 2016. An updated version will be made available once the 2017 paper is released.

## How should I use TSA Past Papers?

TSA Past papers are one the best ways to prepare for the TSA. Careful use of them can dramatically boost your scores in a short period of time. The way you use them will depend on your learning style and how much time you have until the exam date but in general, you should try to do at least 2008 – 2015 once. If time permits, do them twice- practice really does make perfect!

## How should I prepare for the TSA?

Although this is a cliché, the best way to prepare for the exam is to start early – ideally by September at the latest for TSA Oxford and by October for TSA Cambridge. 4 weeks of preparation is usually sufficient for the majority of students. If you're organised, you can follow the schema below:

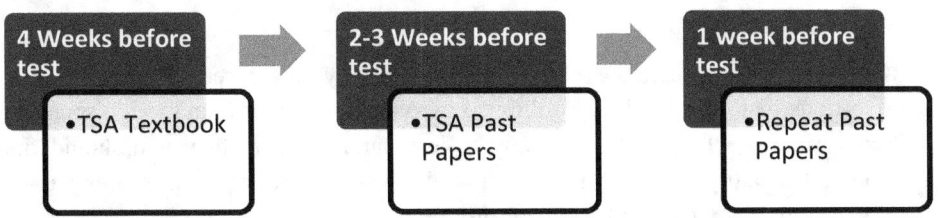

This paradigm allows you to focus your preparation and not 'waste' past papers. In general, aim to get a textbook that has lots of practice questions e.g. *The Ultimate TSA Guide* (**www.uniadmissions.co.uk/tsa-book**) – this allows you to rapidly identify any weaknesses that you might have e.g. identifying flaws, spatial awareness etc.

**You are strongly advised to get a copy of** *'The Ultimate TSA Guide'* **which has 300 practice questions**– you can get a free copy by following the instructions at the back of this book.

Finally, it's then time to move onto past papers. The number of TSA papers you can do will depend on the time you have available but you should try to do each paper at least once. If you have time, repeat each paper (choose a different essay question). Practice really does make perfect!

If you find that you've exhausted all past papers, there are an additional six mock papers available in *TSA Practice Papers* (flick to the back to get a free copy).

## How should I use this book?

This book is designed to accelerate your learning from TSA past papers. Avoid the urge to have this book open alongside a past paper you're seeing for the first time. The TSA is difficult because of the intense time pressure it puts you under – the best way of replicating this is by doing past papers under strict exam conditions (no half measures!). Don't start out by doing past papers (see previous page) as this 'wastes' papers.

Once you've finished, take a break and then mark your answers. Then, review the questions that you got wrong followed by ones which you found tough/spent too much time on. This is the best way to learn and with practice, you should find yourself steadily improving. You should keep a track of your scores on the previous page so you can track your progress.

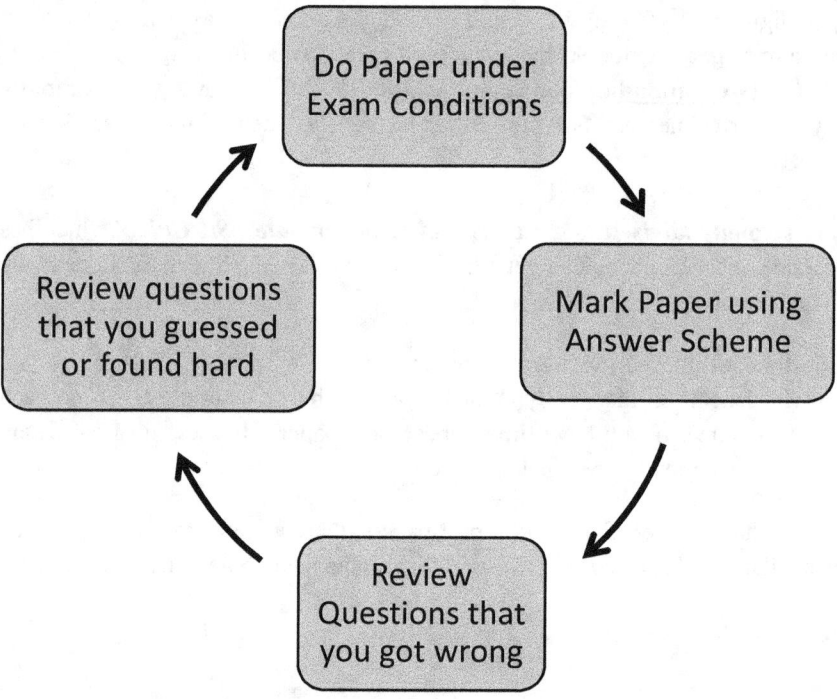

## Scoring Tables

Use these to keep a record of your scores – you can then easily see which paper you should attempt next (always the one with the lowest score).

| Volume One | | |
|---|---|---|
| | | |
| | | |

| Volume Two | | |
|---|---|---|
| | | |
| | | |

## Extra Practice

If you're blessed with a good memory, you might remember the answers to certain questions in the past papers – making it less useful to repeat them again. If you want to tackle extra mock papers which are fully up-to-date then check out *TSA Practice Papers* for **six** full mock papers with worked solutions (flick to the back for a free copy).

| SECTION 1 | 1ˢᵗ Atte | 2ⁿᵈ Attempt | 3ʳᵈ Attempt |
|---|---|---|---|
| | | | |
| | | | |
| Practice Pape | | | |
| | | | |

# 2013

## Section 1

### Question 1: D

The passage is talking about the fact that in Britain, horse meat is a heavily unpopular food substance, and arguing that this is not a logical view.

B) and C) are not valid conclusions from the passage. At no point does it imply that Britain should follow the example of France and Belgium, so C) is incorrect. B) is implied in the passage from the quote from the rival chef, but it is not a conclusion. The reasons given in the argument, if true, do not lead us to think that B) is true. Therefore, B) is not a conclusion.

A), D) and E) are all valid conclusions of this passage. However, the statements in A) and E) go on to support the statement given in D). Therefore, D) is the main conclusion of the argument, whilst A) and E) are intermediate conclusions from the passage.

### Question 2: D

Each thermometer is accurate to within 2 degrees of the temperature stated. The lowest temperature given by any thermometer is 7 degrees, therefore the temperature cannot be any higher than 9 degrees, otherwise this thermometer would have to be stating a temperature more than 2 degrees away from the real temperature.
Similarly, the highest temperature given by any thermometer is 10 degrees, so the lowest that the true temperature could possibly be is 8 degrees. If it were any lower, this thermometer would be out by more than 2 degrees.

The middle thermometer is actually irrelevant in this instance, as the other two thermometers give sufficient information to work out the minimum temperature range.

## SECTION ONE        2013

**Question 3: D**
The passage has described how evidence confirms the predictions of this theory of how the moon was created, and has then incorrectly gone on to conclude that this means we should accept the theory as true. Answer D) correctly identifies the flaw in this reasoning – Just because the predictions of a theory are proven correct does not mean the theory is correct.

B), C) and E) are all irrelevant in this particular situation. The passage is describing how the evidence and facts available are consistent with the theory. However, these 3 answers all describe cases where the facts/evidence are *not* consistent with the theory. Therefore, these are not flaws of the argument, and can be ignored.

A) is completely irrelevant as the passage makes no references whatsoever about how a theory's popularity influences the truth. It simply states that this is the most popular theory.

**Question 4: C**
A) and E) are irrelevant and incorrect as the argument does not make reference to common sense. It claims that the system is flawed, and claims that the original name *Eohippus* was more sensible, but does not claim anything about whether common sense is, or should be, a part of scientific naming conventions. Meanwhile, D) is incorrect because whilst the argument has claimed that *Eohippus* was a more sensible name; it does not claim that the name should be changed back.

B) initially appears to be a valid answer, but on closer inspection we see that B) states that the system is *not the most appropriate*, a claim the passage does not support. The passage does claim that the system is flawed, but does not claim it is not the most appropriate system, so B) is incorrect. However, the passage does claim that a name produce by the current system (*Hyracotherium*) is not the most sensible. Therefore, the passage unarguably supports C) and is thus the correct answer.

## Question 5: A
The spokesperson's argument is that prison is too comfortable, and that this is proven by the large numbers of prisoners reoffending in order to have access to the comforts of the prison. At no point has it been stated that the prisoners would not reoffend if prisons were less comfortable, and we can see that if this is not true the spokesperson's conclusion is no longer valid from their reasoning. Therefore A) is a valid conclusion.

D) is completely irrelevant to the spokesperson's argument, whilst C) does not affect his conclusion that the levels of reoffending prove that prison is too comfortable, so can be ignored. B), meanwhile, actually weakens his argument if true, and is therefore not an assumption. E) actually strengthens the prison officer's argument, if true, so therefore E) is not an assumption either.

## Question 6: D
If 1974 and 1983 are both years in which Arthur's birthdays have been on dates when there were 8 different digits, and then there cannot be a 1 or a 3 in his date and month of birth.

For there not to have been a 1 in his month of birth, it must begin with 0. Additionally, if there is not a 0 (because it has already been used), 1 or 3 in his date and month of birth, the date of his birth must begin with 2. Hence we so far have 2?/0?/1974 and 2?/0?/1983.
Hence his date of birth is either 25/06 or 26/05 as 5 and 6 are the only two digits left. Hence in 1974, the two digits not in the 8 digit format of Arthur's birthday are 3 and 8, and in 1983, the two digits not in the 8 digit format were 4 and 7.
Given that the years are 9 years apart, Arthur must have been 38 in 1974 and 47 in 1983. Hence Arthur must have been born in 1936 (1974 – 38). Hence the answer is D.

## Question 7: D
From the criteria given in the question we can see that in order to qualify for a bonus, a worker must have:
 - An attendance figure of higher than 90% (i.e. absences of less than 10%)
 - An Over Production figure that is not negative (thus either meeting or exceeding targets)
 - A Product accepted figure of greater than 92% (i.e. rejects less than 8% of total output)
We can quickly see now that Smith, Patel and Owololu meet all of these criteria, and that the other workers do not.

## SECTION ONE — 2013

**Question 8: B**
We can follow the local's directions from Akeland. By doing this, we find that Ducton is **south** of Akeland, Cranton is **west** of Ducton and Eksburg is **north** of Cranton. Therefore, we know that Eksburg must be West of Akeland.
However, we do not know how far South of Akeland Ducton is, or how far North of Cranton Eksburg is. Without knowing these relative distances, we cannot know if we need to travel North or South. We only know we need to travel West. Therefore, B) is correct.
Benford is completely irrelevant, and serves to distract you from the real answer.

**Question 9: D**
The argument gives many reasons to support the conclusion that people in rural areas have to travel far to get to pubs and clubs, and have no choice but to drive home afterwards. It then claims that without more pubs/clubs being built in rural areas, the number of people drink-driving will increase. This is much more likely to be true if the population of rural areas will increase in the future, so D) is the correct answer. A) and B) are irrelevant to the argument's conclusion. E) does provide further reasons why people in rural areas may drink and drive, but does not do anything to support the idea that this will increase in the future. C), meanwhile actually weakens the argument, by suggesting that people in rural areas do not choose to drink close to their homes even if pubs/clubs are available.

**Question 10: B**
The passage describes how media personalities employ agents which generate large amounts of income for them, more so than their original activity. It then concludes that talent is no longer being rewarded. Answer B) correctly points out that the original talent may also be rewarded, which is a valid flaw in the passage's reasoning. C), D) and E) are irrelevant and do not affect the conclusion of the passage (that talent is no longer being rewarded). A), meanwhile, actually strengthens the argument by suggesting that personalities without talent can still generate significant income via agents, suggesting that talent is not the decisive factor in profit.

**Question 11: A**
Answer A) correctly identifies the main conclusion to the passage, namely that the government should invest in the film industry.
The other answers are a combination of reasons in the paragraph, and intermediate conclusions which can be drawn from the passage, all of which contribute towards supporting the notion that the Government should invest in the film industry. Therefore, A) is the main conclusion.

## Question 12: B

The starting balance is £0, and each month £50 is paid into the account. Thus, at the end of the first year, the balance will be 12 X £50. Thus, the finishing balance at the end of the first year will be £600.
The average balance for this year is calculated from the average of the starting and finishing balance. Thus the average balance will be (0 + 600)/2 = £300. The interest will be 5% of this balance, thus the interest will be 300X0.05 = £15.
The starting balance for the second year will therefore be £600 + Interest (£15), giving £615. Each month another £50 will be paid in, giving a final balance at the end of the second year of £1215.

Thus the average balance for the second year will be the mean of £615 and £1215. This is £915. The interest paid on this amount will be £915 X 0.05 = 45.75.
Thus, the final balance, once interest for the second year has been paid, will be £1215 + £45.75, resulting in a final balance of £1260.75
To the nearest £10, this is £1260. Thus the answer is B)

## Question 13: B

Marilyn spends 10 hours making 60 cakes, working out as 6 cakes for each hour. Since she charges £6 per hour of time spent, this results in a charge of £1 per cake for the time involved in making them.
Thus each cake costs £1.60 for ingredients, plus £1 for time, resulting in a total cost of £2.60 for each cake.

Thus, each cake would ordinarily be sold at a cost of £4.55 (175% of £2.60).
10% of £4.55 is 45.5p. Thus, friends receive a 45p discount, resulting in a cost of £4.10 per cake for friends.

## Question 14: C

Let X be the % change in wages each time.
If the original change was a decrease, the wage would decrease by X% of the original wage. The next change would be an increase by X% of a smaller amount, thus it would not be as large as the original decrease, and would not raise the wage as high as the original wage.

If the original change was an increase, the wage would raise by X% of the original wage. However, the next change would then be a decrease of X% of a larger amount, thus being larger than the original increase, and bringing the wage *below* the level of the original wage.
Thus, the new wage *must* be smaller than the original wage.

## SECTION ONE  2013

**Question 15: A**

E) is irrelevant to the reasoning of the argument and can be safely ignored. D) and C) are incorrect because the argument does not say whether traditional examinations are fairer, or whether there are any entirely fair ways of assessing students.

B) is incorrect as the argument does not claim or imply that we should only assess A Levels via examinations. It casts doubt on the fairness of assessment by coursework, but this reasoning does **not** necessarily follow on to say that we should only assess A Levels via examinations.

A) correctly identifies a conclusion from this argument, that assessment by coursework is not necessarily fairer than assessment by examination.

**Question 16: C**

The passage argues that factory farming is cruel, and that if people are concerned about animal welfare they should therefore purchase game. At no point is it stated that game meat is not produced via factory farming methods, and if this is not true it weakens the conclusion that people concerned about animal welfare/factory farming should purchase game. Therefore, C) correctly identifies an assumption in the argument.

None of the other possible answers would directly affect the conclusion of the argument if true, and so they are not assumptions.

**Question 17: C**

A) is irrelevant as the argument is discussing waste, not energy consumption. D) is also irrelevant as the fact that councils already have programs to promote washable nappies does not affect the conclusion that the government must do more to promote this. E) is also completely irrelevant.

B), meanwhile, would actually weaken the argument if true, by suggesting that disposable nappies will not cause waste problems, as they are compostable. Only C) would strengthen the conclusion that disposable nappies cause significant issues of waste, and washable nappies must be promoted instead.

## SECTION ONE  2013

**Question 18: C**
The car travels 3 times as fast as Sven on his cycle. Therefore, without the breakdown Helga should take 10 minutes to arrive. She travelled two thirds of the way before the car broke down, so she would have been travelling for 6 minutes and 40 seconds before the breakdown.
Helga then had to travel the final third of the journey on foot, travelling at 1/3 of the speed of Sven's cycle. Therefore it would take Helga 90 minutes to walk the full journey, and would take her 30 minutes to walk the final third of the journey.
Therefore Helga would take 36 minutes 40 seconds to arrive. This is 6 minutes 40 seconds longer than Sven (who takes 30 minutes as stated in the question). To the nearest minute this is 7 minutes.

**Question 19: E**
The only way it is possible to make a hexagon from one straight cut across the cloth is by starting the incision in either the left-most side (labelled A below) or the right-most side (labelled B below), and cutting to the **nearest** of the 2 sides labelled X (i.e. along one of the two dotted lines). As can be seen, these cuts will result in a triangle, and not a quadrilateral. We can therefore see that it is impossible to produce a hexagon and a quadrilateral from one straight cut in this cloth.

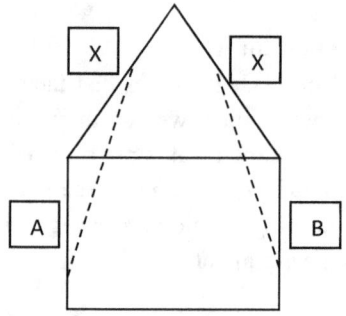

**Question 20: E**
The question states that Thomas is only free on mornings, so all afternoon sittings can be ignored. We are only considering morning tests.
He wishes to take the test in the main hall, so is looking for a test which has more than 30 candidates sitting it. The number of candidates currently booked onto a given test will be 100 minus the number of spaces currently available. Thus, if there are more than 70 spaces available, there are currently less than 30 people taking the test, and it may not take in the main hall. Therefore, Thomas will only book a test which currently has fewer than 70 spaces available. As we can see from the list, all tests except Monday currently have fewer than 70 spaces available, so we should consider all tests except Monday.
However, he also wants to be with the smallest number of other candidates possible. This will result from booking onto a test which currently has the largest number of spaces available (as long as it is less than 70). As we can see, from the tests we are considering, Friday has the largest number of spaces available.
Therefore, Thomas should book a test for Friday.

**Question 21: E**
A) and D) are completely irrelevant statements.
B) does not necessarily strengthen the argument as the fact that politicians have high-profile affairs does not necessarily support the notion that the personal aspects of politicians' lives should be reported.
C) actually weakens the argument, suggesting that there are negative consequences to reporting on politicians' private lives.
E) strengthens the argument, if true, in that it describes how politicians claim to be honest in aspects of their lives. Thus, by reporting on the private aspects of a politician's life, the media can provide the public with important insights into whether these claims of honesty are genuine, and thus provide important aspects into the politicians' values. This ties in with the argument's reasoning.

**Question 22: B**
The argument's reasoning can be summarised as "we can conclude that A is happening because B happened and we know C." In the passage, "A" is a person being a criminal, "B" is a person living a lifestyle that they cannot legally afford, and "C" is that people living a lifestyle they cannot legally afford are often criminals.
Answer B) follows the same pattern of reasoning. In this instance, "A" is someone spending more than half an hour on a piece of work, "B" is writing more than 4 pages, and "C" is the fact that it is impossible to write more than 4 pages in half an hour. Therefore, when "B" happens and someone produces more than 4 pages, we know "C", so we can conclude "A".
D) is the answer which most appears to follow this reasoning on first glance.

However, when we assess answer D), we find that the reasoning can be summarised as "A is often followed by B. A is happening, so we can expect B to happen soon". Here "A" is a person saving up, and "B" is that person going on holiday. This is not the same pattern of reasoning as in the question.
C)'s reasoning can be summarised as "For A to happen, B must have happened. A happens, so we know B must have happened". Here, "A" is the bank being robbed, and "B" is very careful planning, with attention to detail.
E) follows similar, but slightly different reasoning to C). E) can be summarised as "For A to happen, B must happen. We want A to happen, so we do B". Here, "A" is the police catching criminals, and "B" is watching how people spend their money.
Meanwhile, A) follows a completely irrelevant style of reasoning, simply concluding that one possible reason for an event has not occurred, and trying to find other possible explanations.

## Question 23: C
The principle of the question is that those who are best at a given task should be rewarded for it. Therefore we look for an answer which follows this principle.

A) and D) are incorrect as these both refer to a situation where people are rewarded for *how much* work they do, rather than *how good* they are at a given task. E) is incorrect as it essentially reverses the principle by deciding who is best based on how much money they make, rather than basing how much money someone makes on how good they are. B) is incorrect as it refers to eliminating poorly performing people, rather than rewarding high-performers.

C) follows the principle. It identifies an important aspect of work, then says that someone was picked out for being best at this aspect, and rewarded for it. This follows the principle used in the passage.

## Question 24: C
The showroom is 24m by 12 m, resulting in a total area of $288m^2$. The showroom thus has 2 edges 12m long and 2 edges 24m long, so there is 72m of tape required to cover the edges of the showroom.
Each rectangle of carpet is 8m by 4m, so has a total area of $32m^2$. Thus, we can see that there will be 9 rectangles of carpet required to be taped along all seams.
Each rectangle of carpet has 2 sides 8m long, and 2 sides 4m long. Thus, each rectangle will require 24m of tape to cover all the sides of a given rectangle of carpet.
Thus, 216m of tape (24 X 9) would be required to cover all the edges of each rectangle of carpet. However, less tape than this is required because:
- The points where the carpet joins the edges of the showroom have already been accounted for, and do not need to be covered again.
- Each seam will only need to be covered once.

Thus, to calculate the amount of tape required to cover the seams between carpets:
- First we subtract 72m from 216m, to give us the length of tape required just to cover the points when carpet meets carpet. This gives us 144m.
- Then we divide this number by 2, as each seam only needs to be covered once. We do not need to apply tape to both bits of carpet in a given seam. This leaves 72m of tape required.

Thus, we can see that 72m of tape is required to cover the seams between carpets, and then another 72m is required to cover the edges of the showroom. Thus, 144m of tape is required overall.

# SECTION ONE 2013

**Question 25: B**
The show runs for 3 full weeks, plus one additional Saturday. Each week has the following performances:
- 6 Nightly performances between Sunday and Friday
- 2 Nightly performances on Saturday
- 3 additional Matinee performances

Thus, there will be 11 performances each week, resulting in 33 performances overall, just considering the number of weeks. Then we need to factor in the additional Saturday, which will have 3 performances, giving 36 performances
Then we need to account for the closure of the theatre on 25$^{th}$ December. If the 28$^{th}$ December is a Saturday, the 25$^{th}$ will be a Wednesday. Thus it would have had 1 night performance and 1 matinee performance, so we need to subtract 2 from the total number of performances.
This gives a final total of 34 performances during the run.

**Question 26: E**
Answers A) through D) all follow the same pattern as the question. Each thing in question can be paired with each of the others, giving 30. This counts each combination twice, so we half this number to get 15 as the total number of combinations possible.

However this does not apply to 6 friends sending each other Christmas cards. This time, each friend sends 5 cards, so we still multiply 6 by 5 to get 30. However, this time we do not half the number, as each friend will both send a car to each other friend, and receive one in return. Thus, person 1 will send a card to person 2, but will also receive a card from person 2.

Therefore, we do not half the number, as we do wish to consider each pairing twice. Therefore the total number of cards sent will be 30, 5 per friend.

**Question 27: D**
The argument begins by describing how a belief that family planning services can solve the problem of overpopulation is Naïve, and then goes on to give reasons why it can in fact only be properly tackled by economic development. Thus, the main conclusion of the passage is that given in D).

E) is actually a reason in the passage, which helps support this main conclusion. B) and C) also both go on to support the conclusion given in D).
A) is an irrelevant point.

## Question 28: E

The argument's structure can be summarised as "If A could happen/was happening, B would happen/would have happened by now. B has not happened, so A cannot be happening".

Answer E) follows this structure, claiming if intelligent life existed, we would have evidence for it. We do not have any evidence for it, so there is no intelligent life.

From the other possible answers, C) is the closest. C) argues as "If A happened B would not happen. A hasn't happened, so B is happening". This is not the same structure as discerning that A has/has not happened on the basis of B happening/not happening.

A) reasons as "We know that A cannot happen because it has not happened outside of fiction", which is not the same reasoning as in the passage.

B) argues as "If A happened, B would happen. B has happened, so A must have happened". Meanwhile D) argues as "For A to happen, B must happen. B happens, therefore A must have happened". Both these arguments are incorrect, and neither follow the same structure as the passage.

## Question 29: A

C) and D) are irrelevant statements, whilst E) actually contradicts the argument's conclusion, which concludes that video is as much a medium for great art as any other form of expression.

B) is not an assumption because the question explicitly states "if they [the jury] are right". Thus, it has not assumed that the jury are right, and is simply talking about a case which would be true *if* they are.

However, the question does **not** state that work with emotional force and complexity is necessarily capable of being great art. The emotional force and complexity of the video medium described by the jury is the main reason given to support the conclusion that this medium is capable of being great art. Therefore, if the statement in A) is not true, the argument's reasons no longer necessarily lead on to its conclusion. Thus, A) is an assumption in the passage.

## Question 30: B

Dividing 20,000 by 40 gives us the number of gallons of fuel mixture required to drive 20,000 miles. However, since the mixture is only 75% ethanol, we must then multiply this number by 0.75 (i.e. 75/100) to calculate the number of gallons of ethanol required for this distance.

We must then divide the number of gallons required by the number of gallons 1 field of sugar beet produces (550) in order to get the final answer.

B) correctly follows this working, producing the correct answer of 0.68

# SECTION ONE 2013

**Question 31: B**
We can work out the distance each car can travel on each tank of petrol and then select the furthest.
- Clipper: 12 miles per litre, 60 litres per tank, = 720 miles per tank
- Ghia: 11 miles per litre, 70 litres per tank, = 770 miles per tank
- Sedan: 10 miles per litre, 75 litres per tank, = 750 miles per tank
- Estate: 8 miles per litre, 80 litres per tank, = 640 miles per tank
- Saloon: 5 miles per litre, 82 litres per tank, = 410 miles per tank

The Ghia travels furthest on one tank at 770 miles, hence the answer is B.

**Question 32: B**
A) cannot be concluded because we do not know how much of the population comes from each category, so we cannot conclude whether the overall % of people under 55 with no natural teeth is >50. Similarly E) cannot be reliably concluded without this information.

C) and D) cannot be reliably concluded as we cannot make any conclusions about the causes of tooth loss, or the effects of tooth loss on employment prospects, from this data. Any such inferences would be a confusion of cause and correlation.
B) can be reliably concluded as the % of Professionals, Employers and Managers with no natural teeth is lower than the % of semi-skilled and unskilled manual workers in the same age group with no natural teeth.

**Question 33: C**
The argument concludes that organic sales will only continue to rise if there is a nutritional benefit from eating organic food. However, C) correctly points out that organic sales have risen for some time now, supposedly without there being any nutritional benefit from eating organic food. Thus, C) presents the strongest challenge to the article.

A) does not necessarily present a challenge to the argument. It implies that the article in the independent has not prevented people from buying organic food, but this does not necessarily mean that a lack of nutritional benefit will not stop organic sales rising. It could be that people who read the independent are inherently more likely to buy organic food even without a nutritional benefit.

D) and E) are irrelevant. B) is relevant to the argument but does not affect the conclusion as it suggests the nutritional benefit will not affect whether people choose to purchase organic good.

# SECTION ONE 2013

**Question 34: B**
The argument concludes that if a Swiss political system were adopted in Britain, the populace would be happier. B) correctly identifies that this assumes the only reason Swiss people are happier is due to their political system. If this is not true, then we cannot state that adopting a Swiss political system *would* make the British populace happier so the argument's conclusion is invalid.

A), C) and E) do not affect the conclusion so are not assumptions, and can be ignored. D) is not a valid assumption because the argument does not assume there is only one reason why people are disillusioned with politics. Even if there are many reasons, this does not mean that removing one of them will not make the populace happier. Thus D) is not *required* to be true in order for the conclusion to be valid.

**Question 35: B**
A) is an irrelevant statement, whilst C) and D) are reasons given in the passage to help support the main conclusion.

E) is an intermediate conclusion, which helps support the main conclusion, which is the statement given in B). The statement in E) supports the statement in B), but the statement in B) does not readily support the statement in E). Therefore we can see that E) is an intermediate conclusion and B) is the main conclusion of the argument.

**Question 36: B**
After 16 days, the average was just over 6 miles. 6 miles per day for 16 miles is 96 miles. For the 9 days after 16 days, 2 miles were run each day and so after 25 days, the total miles run is 96+18=114 in 25 days.
- If I run 8 miles the day after, this will be 122 miles in 26 days which is less than 5 miles per day.
- If I run 8 miles for 2 days after, this will be 130 miles in 27 days which is less than 5 miles per day.
- If I run 8 miles for 3 days after, this will be 138 miles in 28 days which is less than 5 miles per day.
- If I run 8 miles for 4 days after, this will be 146 miles in 29 days which is more than 5 miles per day.
- Hence it will take 4 days until the average can be brought back up to 5 miles on average per day.

## Question 37: C

The maximum number of times it can be activated will assume it always plays the shortest tunes possible. If it plays 3 different tunes, the shortest 3 it can play will be 10+15+15 seconds long which is 40 seconds.

In an hour, there are 3600 seconds. 40 goes into 3600 90 times, hence the bird can play 3 different tunes 90 times in the space of an hour. Hence the answer is C.

## Question 38: E

The question has stated that the number of visitors *per hour* is higher for Tuesday and Friday than for the other days, yet less customers were received on Tuesday than any other day except Sunday, and on Friday the number of customers was the same as Monday, Thursday and Saturday. Therefore, both Tuesday and Friday must have had shorter opening hours than these days, so A), C) and D) are incorrect.

We can also see that Tuesday had ¼ less visitors in the day than Monday, Thursday and Friday. Since it had more visitors per hour, we can conclude that the opening hours must have been more than ¼ less than for Monday, Thursday and Friday. These days had 10 hours open time in total. Therefore, on Tuesday the shop must have been open for less than 7.5 hours.
The only possible answer which fits in with all of these conclusions is E)

## Question 39: D

The passage's main conclusion is that the USA's actions in the Haiti crisis proves it is the only genuine superpower. The statement in D), if true, would weaken the argument as it suggests that the EU has a much greater military force than was seen in the Haiti earthquake crisis, but they were not used because they are not an integrated force that is mobilised for humanitarian crises. This implies that they may possess a force comparable to the United States that could react if necessary, and thus class as a superpower.

A) is an irrelevant statement whilst E) does not affect the argument's conclusion.
B) does not affect the argument's conclusion as simply having as many men in the military does not necessarily mean that the EU is a comparable superpower to the USA.
C), meanwhile, would actually strengthen the argument's conclusion by suggesting that the EU's ability to respond is limited, and thus it is not a comparable superpower to the USA.

## Question 40: D

A) is contradictory to a statement in the passage because it is stated that no-go areas for trawlers would enable fish to reproduce safely, which carries an inherent implication that fish cannot currently reproduce safely. B) is irrelevant as the passage simply states it should be top priority, not that current coverage is lacking. C) is irrelevant to the conclusion, and also contradicts directly a statement in the passage.

E) is not an assumption as the passage refers to no-go areas for trawlers of *any nationality*. Thus the idea of "foreign trawlers" being responsible for dwindling fish stocks is not required to be true for the argument's conclusion to be valid.

Only D) is required for the conclusion to be valid, yet it is not stated at any point that Fish farming is impossible without wild fish as food. It is stated that fish farming is "futile" without the wild fish, but not that it is impossible.

## Question 41: C

A) and E) are irrelevant to the argument's conclusion, as they say nothing about whether investing in irrigation schemes would help poor countries to feed their populations. D) is completely irrelevant to the notion of solving world hunger.

B) meanwhile actually weakens the argument's conclusion, suggesting that lack of mechanisation is not the main cause of malnutrition in poor countries and that wars/natural disasters are major causes, which would not be solved by mechanisation.

C) suggests that currently much of the excess food cannot be used to provide for poor countries, and in doing so it strengthens the argument that investing in mechanisation for poor countries would be a better solution.

## Question 42: C

The graph shows us that each application adds value to the crop as follows:
- The first application adds £500 value to the crop.
- The second application adds a further £2000 of value to the crop (giving a total of £2500 extra value, once we add in the value for the first application).
- The third application adds a further £1500 of value to the crop (giving a total of £4000 extra value, once we include the value added from the first two applications).

We can immediately see that each additional application after the third adds only £500 extra value, and will cost £1000 per application, so after the third application, adding another application will reduce profit. Hence, D) and E) can be ignored, as these numbers of applications would not maximise profit.

We can see that 2 applications will add £2500 extra value, and will cost £2000, resulting in £500 total profit from 2 applications.

We can see that 3 applications will add £4000 extra value, and will cost £3000, so will result in £1000 profit from 3 applications. Thus, 3 applications is the most profitable.

**Question 43: C**
The passage is asking about a train from Teovil to Erd, departing at 20 minutes past 10. All other trains mentioned in the question can be ignored.
- Ordinarily the train would arrive at Erd 51 minutes after departure, so the expected arrival time is 11:11am
- The quickest way to calculate how many minutes late the train would be is to calculate how much delay has been encountered.
- The train has arrived in Uble station on time, so any time taken to reach Uble is irrelevant to how late the train is.

First, calculate how long the train should take to then travel on to Erd:
- The train would ordinarily wait 3 minutes in Uble station.
- It would then travel to Erd, covering a total distance of 24 km, as seen from the distances chart in the question. If the train were on time, it would cover this distance at 60km/h, thus taking 24 minutes to travel this distance.
- The train also needs to make stops at Ergen, Lowley and Aregon stations. Ordinarily it would wait 3 minutes at each station, giving a total of 9 minutes spent waiting at stations.

Thus, ordinarily there would be (3+24+9) = 36 minutes between the train arriving at Uble, and the train arriving at Erd.

Now calculate how long the train actually takes to make this journey:
- The train sits in Uble station for 22 minutes.
- The train would then cover the 24km to Erd station. Since it is late, it would now travel at 80km/h, a third faster than the train would normally travel. Therefore, it would cover the distance in ¾ of the time it would normally take, and would cover 24km in 18 minutes.
- The train would also need to stop at Ergen, Lowley and Aregon stations as usual. However, since it is late, it would only stop for 2 minutes at each station, giving a total of 6 minutes spent waiting at stations.

Thus, thanks to the delay there will be (22+18+6) = 46 minutes between the train arriving at Uble and the train arriving at Erd. Thus the train will arrive at Erd 10 minutes late.

**Question 44: E**
Answer A), the view from the East, shows a door in an East-facing wall. We have been told that the only doors are in the westernmost wall (which will thus only be viewable from the East), and in a North-facing wall (which will thus only be viewable from the North). Thus, A) is incorrect.
B), the view from the West, is not a possible view because there are no doors in sight. We have been told there is a door in the westernmost wall, and as we can see from the top-down view, the westernmost wall will be viewable from the West. Thus, there must be a door viewable from the West, so B) is incorrect.
C), a view from the south, is incorrect as there are only 2 windows in the image. We have been told there are 3 south-facing windows, which must be viewable from the south. Therefore, C) is incorrect.

D) shows the correct number of South-facing windows, but it is not a possible view because the layout of the church is wrong. As we can see from the top down view, the longer section of the church extends to the west side of the church. Thus, when viewed from the south, the longer section of the church would be on the left hand side. In D), the longer section is on the right-hand side, so this is not a possible view of the church.
E), a view from the North, is a possible view. The longer section of the church we discussed in D) should proceed to the right-hand side of the church when viewed from the North, which is what is seen in view E). Also, we are told that there are 2 north-facing windows and a door in a North facing wall. Thus, we should be able to see 2 windows and a door when viewing the church from the North. We see these features in view E).
Thus, E) is a possible view.

**Question 45: A**
B) is a fact stated in the argument, and is not a conclusion, because it is not supported by any other reasons given in the argument.
C), E) and A) are all valid conclusions that can be drawn from the argument. When we examine them further, we see that C) and E), if true, would support the statement given in A). Therefore, C) and E) are intermediate conclusions, and A) is the main conclusion of this argument.

D) can be treated as a conclusion in the argument, which can be drawn from the argument's description of how communities can exert control over behaviour. However, it is not the main conclusion, because it is only supported by a small section of the argument, whereas the main conclusion should be supported by the argument as a whole.

**SECTION ONE**     **2013**

**Question 46: D**
The reasoning in the passage can be described as follows: "A always happens if B happens. A doesn't happen, so B can't have happened." Only D) follows this reasoning (with "A" being John enjoying the movie, and "B" being there being a big star in the movie).

C) reasons as "A happens if B happens. B happens, therefore A will happen". A) reasons as "If A happens, B will happen. A happens, therefore B will happen" (although the second part is stated in an inverse way). B) reasons as "If A happens, B happens. B happens, so A must have happened". B) is incorrect reasoning, and none of these follow the same pattern as the passage.

E) meanwhile simply describes a situation and states it happens frequently.

**Question 47: C**
The passage describes a principle of rewarding people for good performance, rather than as a matter of course. Only C) follows this principle, describing Tim getting a bicycle as a reward for his high grade.

B) is a different situation describing punishment for those performing poorly. A) is a direct opposite, describing how all employees get money because the company is in profit, rather than as a reward for their own performance. D) also describes a situation where everybody gets a financial reward, regardless of personal performance. E) meanwhile does not describe a reward specifically being given, it simply describes a positive situation resulting from good performance.

**Question 48: B**
Let the value of a Red note be termed "R", a Green coin be "G" and a Blue coin be "B".

The question states that the smallest denomination is worth 1k, and that each higher denomination is a whole number multiple of a smaller denomination. Thus, we know that the value of each denomination is a positive integer of "k". Thus, in all equations with a certain number of "k", the "k" can safely be discarded.

Hence from the amount of change you get when paying for something worth 135K with a red note, we can write the equation R-3G-B=135. Rearranging this, we get R=135+3G+B

We also know that something worth 33K is paid for with 4 green and 1 blue coins. Hence 33=4G+B

The question states that one of the denominations is equivalent to 1k, and we know R is more than 135k, so we know that either a blue coin or a green coin must be equivalent to 1k.

- If B=1, we can see that 4G=32 and so G=8. Hence from the equation R=135+3G+B we see that R=160
- If G=1, we can see that B=33-4G=29. Hence from the equation R=135+3G+B we see that R=167

Each higher denomination is a whole number multiple of the lower denominations. 167 is not a multiple of 29, so the answer cannot be G=1, B=29, R=167. 160 is a multiple of both 1 and 8, so B=1, G=8, R=160 is the only possible answer.

Hence B=1, G=8, R=160 is the answer. G=8 and R=160 hence 20 green coins are worth 1 red coin. Therefore the answer is B.

# SECTION ONE — 2013

**Question 49: A**

The total number of votes in favour of the motion is 5, whilst the total number of votes against the motion is 8. Thus, there are currently 3 more votes against the motion than for the motion.

If 2 committee members who previously voted no decide to vote yes, and the same committee members abstain, then the total number of votes in favour of the motion will be 7, and the total number of votes against the motion will be 6.
Therefore, the minimum number of committee members who need to change their vote for the motion to be passed is 2. Hence, the Answer is A)

**Question 50: E**

Andrew *may* be able to calculate the number of houses in the street by knowing Amy's house number and that opposite, but *only* if the houses in the street are numbered consecutively, starting at one end of the street, going up to the end and back down the other side. Then he will be able to work out how many houses are either side of Amy's house. However, if the houses in the street are numbered with the odd numbers on one side and the even numbers on the other side, he cannot work out how many houses have a larger house number than Amy's house. Therefore, A) and B) are incorrect.

C) is also incorrect because if the house opposite Amy's is 26, then the houses are numbers with even houses on one side and odd numbers on the other, so he cannot work out how many houses there are. In fact, he can work out how many houses there are only if the house opposite Amy's house **is not** number 26.

Similarly, D) is incorrect because if the houses are numbered with odd numbers on one side and even numbers the other, knowing the numbers either side of Amy's house will not allow him to work out the number of houses on the street.

However, Andrew can work out the numbers of the houses either side of Amy's, as follows:
- If the house opposite Amy's house is number 26, then we know the houses are numbered odd numbers one side, even numbers the other. Therefore, the houses either side of Amy's will be 23 and 27
- If the house opposite Amy's house is **not** number 26, then we know that the houses must be numbered consecutively, going up one side and back down the other. Therefore, we know that the houses either side of Amy are 24 and 26.

**END OF SECTION**

# Section 2

*Can you ever know whether anyone else has thoughts and feelings like yours?*

## *Introduction*

➢ This is a complex philosophical question based on an empirical issue. In order to answer this question, which I would not advise unless you have a solid and wide philosophical background, it is first necessary to explore what is meant by knowledge. Explore and define the term in as clear a manner as possible. One such definition could be: 'facts, information, and skills acquired through experience or education; the theoretical or practical understanding of a subject.'

➢ However, this is definition is largely scientific in nature. While there is a large scientific element as explained below, it is important not to ignore the mention of 'thoughts' and 'feelings', which it could be argued go beyond the scientific realm.

## *Philosophical theory*

➢ It is a good idea to explore a number of different epistemological approaches, if possible backed up by various philosophers such as Plato, Descartes etc. However, if you do include such thinkers, make sure you keep their arguments simple enough to be referred to throughout the essay, but not so simple as to misunderstand their arguments.

➢ Ensure you are confident with a thinker if you want to include them in your essay – it is better not to ascribe an idea to a thinker than to ascribe a false or erroneous one.

➢ Another area to explore would be the difference between being able to know 'thoughts or feelings' as opposed to other forms of knowledge, such as basic predictable instincts.

➢ Feelings for example have a range of levels of complexity – for example I can say with quite a lot of certainty that if you put your hand in a bowl of boiling water you will feel pain, but determining other, more complex feelings such as love or jealous are much harder to see.

## Biological arguments
➢ While that our biological apparatus for sensing the world is fairly similar from person to person, that sensing process cannot in practice be decoupled from the processes of attention filtering and emotional interpretation, which are likely to vary widely between people: an interesting facet of this question is the role that emotions play in our perception of the world. Scientific understanding of how this works is still at a basic level, but experiments are showing that a change in emotional state can often affect one's perception. And of course the sense data we receive from the world has to pass through the filter of our attention as well, and this filter is highly sensitive to emotional context. All of this explains why, if we have a strong emotional association with a particular colour, taste, smell, sound or texture, we start to observe it more often (and possibly differently) in the world.

## Nature vs. nurture
➢ You could use the Nature vs. Nurture debate (whether a person's development is predisposed in his DNA, or a majority of it is influenced by this life experiences and his environment) - even though the general wiring of the human brain is specified by genetics, a great deal is influenced by nurture. For example, in the visual system, each cell differentiates so as to handle only signals coming from one eye, and this is unique to every individual.

## Conclusion
➢ Summarize the main points made on each side of the argument in the essay.
➢ You may wish to come to a decision either way, or it is equally fine to sit somewhere in the middle, so long as this is fairly justified

## Should the supply and use of all drugs be legalised?

### Introduction
- This essay is relatively simple to answer if structured correctly due to the wide availability of arguments both for and against drugs. However, it is important that the arguments you use are measured and focused in approach – merely writing down every single argument you can think of connected to narcotics will not lead to a good essay as you will fail to achieve sufficient depth or clarity of thought.
- This essay requires a clean and clear definition of drugs. Drugs can be defined as 'a medicine or other substance which has a physiological effect when ingested or otherwise introduced into the body.'
- Note that this term has a very open ended definition, and can be seen include a number of substances which would not conventionally be considered as a drug – such as alcohol, or caffeine for example.
- An important issue to note that not all drugs are the same in terms of the detrimental effect they have on the body- for example, some drugs such as caffeine have a negligible effect on one's health.

### Arguments against
- A key issue at the heart of this question is liberty. One argument could be – it's my body, so why should I let the laws of the government constrain what I do with it? I could just as easily damage my body by sticking a metal fork into a plug socket as I could by taking an illegal substance. Any law which limits what one can do to their own body can therefore be seen as an affront to their individual liberty.
- Another issue is that not all drugs are illegal – for example alcohol and tobacco are drugs which, despite having a large negative impact on one's health, remain legal (albeit not available to those under the age of 18)

### Arguments for:
- However, the counter argument is that of paternalism – that the government prevents the supply of drugs because it is in the best interests of an individual's liberty and wellbeing to do so – if they were to allow unlimited access to very addictive substances such as heroin, individuals would quickly become addicted and would be unable to stop taking the drug – leading to their eventual death.
- Eliminate the criminal market place - The market for drugs is demand-led and millions of people demand illegal drugs. Making the production, supply and use of some drugs illegal creates a vacuum into which organised crime moves. The profits are worth billions of pounds. Legalisation forces organised crime from the drugs trade, starves them of income and enables us to regulate and

- control the market (i.e. prescription, licensing, laws on sales to minors, advertising regulations etc.)
- ➤ The price of illegal drugs is determined by a demand-led, unregulated market. Using illegal drugs is very expensive. This means that some dependent users resort to stealing to raise funds. Most of the violence associated with illegal drug dealing is caused by its illegality. Legalisation would enable us to regulate the market, determine a much lower price and remove users need to raise funds through crime. Our legal system would be freed up and our prison population dramatically reduced, saving billions. Because of the low price, cigarette smokers do not have to steal to support their habits. There is also no violence associated with the legal tobacco market.
- ➤ Safety - prohibition has led to the stigmatisation and marginalisation of drug users. Countries that operate ultra-prohibitionist policies have very high rates of HIV infection amongst injecting users. Hepatitis C rates amongst users in the UK are increasing substantially. In the UK in the '80's clean needles for injecting users and safer sex education for young people were made available in response to fears of HIV. Harm reduction policies are in direct opposition to prohibitionist laws.

*Conclusion*
- ➤ An important distinction in the question is that it refers only to the supply of drugs – what could you say about the illegality of the demand of drugs?
- ➤ If you have a more nuanced argument, like the example given in the introduction, this should be restated here with the arguments that justify it.
- ➤ Restate your position and summarise the arguments and counterarguments you've presented above.

## Do countries benefit from immigration?

### Introduction

- Like with the drugs question above, this question is easy to answer due to the abundance of arguments on both sides, however, again, it is vital that arguments are deployed carefully in a sophisticated manner in order to create a convincing case one way or the other. It is therefore very important with this question to have a clear and concise opening paragraph in which you introduce the reader to the arguments that you are about to make. As well as framing the essay well, this has the additional benefit of helping you keep the rest of the essay structured.
- To answer this question you must define two terms – immigration, which is simple (the action of coming to live permanently in a foreign country) and benefit – which is harder to define.
- There are a number of different potential areas which immigration could be seen to be beneficial – economically, socially, culturally, or politically, to name a few. Pick a couple of these and focus your argument on these areas, but make sure you make it clear this is what you are doing.

### Argument

If for example you take economic, you would need to assess the costs and the benefits of immigration.

- For example: benefit – immigrants have skills, cost: immigrants might use up resources such as social security.
- Another example, culture: benefit – a large number of immigrants will help create a vibrant and multicultural society, cost: cultural integration might be hard to achieve and will lead to friction between different communities.

A good distinction to make is between the impacts on host countries, and of origin countries. There are numerous arguments which can be used for either side.

To give a brief description of the possible arguments available:

**Host countries:**

*Positive*
- Job vacancies and skills gaps can be filled by immigrants, which will improve the competitiveness of the host country economy. This, in turn can lead to sustained economic growth which is maintained through the increase in the supply of labour provided by the immigrants.
- Social security can be improved by the contributions of new workers and they also pay taxes.
- Immigrants bring energy and innovation to a country's economic sphere.
- Host countries are enriched by cultural diversity brought by immigrants.

*Negative*
- There may be a depression of wages and employment as immigrants replace natives in the workforce.
- Having workers willing to work for relatively low pay may allow employers to ignore productivity, training and innovation, and migrants may be exploited.
- Increases in population can put pressure on public services, such as the NHS for example.
- There may be cultural integration difficulties and friction with local people.

**Origin countries:**

*Positive*
- Developing countries benefit from remittances (payments sent home by migrants) that now often outstrip foreign aid.
- Unemployment is reduced and young migrants enhance their life prospects.
- Returning migrants bring savings, skills and international contacts.

*Negative*
- Economic disadvantage through the loss of young workers
- Loss of highly trained people, especially health workers
- Social problems for children left behind or growing up without a wider family circle

*Conclusion*
- Summarize the main points made on each side of the argument in the essay.
- This is a very open ended question – ensure your answer reflects this.
- Be sure to link your conclusion back to the question by referring to immigration through a cost/benefit analysis.

*How should we evaluate advances in science?*

This is a difficult question in that it requires an in depth knowledge of both philosophy and science, and the interaction between them. It can be made simpler by clarifying in your opening paragraph the manner in which you are going to make the argument – i.e. from the points of view of profit and of societal benefit. Never the less, it remains a challenging question.

*Introduction*
- There are two main ways to evaluate advances in science – either through the amount of people the scientific advance helps in regard to lives saved or some other metric, or through the amount of profit that it creates for the scientist/company who invented it.
- From a moral point of view, the importance of a scientific advance is clearly the former – how it helps improve people's lives.
- From an economic, and possibly a more realistic point of view however, the reality might be different. In our capitalist society, profit is a huge driver of scientific advances. Whether this is a good thing or not is a good way of addressing wider political issues while answering this question.
- A key issue at the heart of this question concerns what the purpose of science is – should it be to improve lives or simply to make profit?

*Assessing profit in science*
- Due to the involvement of private, profit making corporations in the scientific field, you could argue that this is no longer the most important. A firm which creates scientific advances which make no money will quickly go bust and stop producing advances. This is an issue that you could explore morally – should this be the case? What are the implications for society? What are the implications for the future of scientific research?
- You could also explore the implications for having research publicly funded versus research which is privately funded. You could argue that private research funding is more likely to flow to issues which generate the most money rather than what is best for the people, leading to advances in areas such as arms and weaponry which might be at the detriment to society as a whole.

## SECTION TWO  2013

*Counter argument*

The other side to this argument is that you could argue this is a false dichotomy, and that given the nature of scientific research industry, any invention that helps people sufficiently will make money anyway. Should this be the case? Try and assess the issue from a moral rather than a purely scientific/profit-based viewpoint.

*Time lag*

Another angle of assessing the profit-based view of science is through timeframe – an advance might not be immediately profitable or beneficial, and it might take many years before an advance is recognised as being useful. An argument therefore could be made that focuses more on an advancement's long term impact rather than immediate impact.

*Conclusion*

- Make sure that you include potential counter arguments for any of the above points that you make – it is vital that your essay comes across as balanced and not one-sided.
- In your conclusion you could argue that this distinction is not absolute – for example a scientific advance could both save lives and make profit for a company.
- Another distinction you could make in your conclusion is between how we ought to evaluate advances, and how they are actually evaluated. How could the current situation in regard to the profit motive be improved? Could, for example the government introduce legislation to prevent the negative effects described above? If so, what kind of laws would they be? Exploring these political implications is a good way to end the question.

**END OF PAPER**

# 2014

## Section 1

**Question 1: C**

The passage discusses reasons why spending money on trying to increase adult participation in sport is futile, citing research showing that it is what people do in their childhood that influences their sporting habits later in life. It them claims that schools should be given money to increase sporting participation amongst children. If we accept the other reasons in the paragraph as true, we have good cause to believe this claim. Thus, answer C) is the main conclusion of this passage. Answers A), B) and E) are all reasons given in the passage, which contribute towards supporting C). Thus, they are not main conclusions.

Answer D) is irrelevant, the passage is not discussing the technicalities of how schools are able to implement sport, it is simply discussing whether they should be given money to do so.

**Question 2: C**

The longest calendar months have 31 days in total. This is 4 weeks (28 days = 4 weeks), with 3 days leftover. The maximum number of working days will occur if the month starts on a Monday. Counting from this day, the $4^{th}$ week ends on a Sunday, and thus the 3 extra days would be Monday, Tuesday and Wednesday, which are all working days.

The total number of working days can thus be calculated:
- Monday to Friday for 4 weeks is a total of 20 days
- Alternate Saturdays are worked, so there would be 2 Saturdays
- Adding in the extra 3 days after the $4^{th}$ week ends, this gives a total of 25

Thus, there are 25 working days. We are told the neighbours alternate driving, so in this example, one will drive 12 days and the other will drive 13 days.

## Question 3: C

This passage discusses how traditional book sales are declining whilst e-book sales are soaring, and concludes by saying that this means that this means e-books have attracted readers away from paper books and towards digital copies. Answer C) correctly identifies that this conclusion is not valid – e-books could have attracted new readers who never read before, and printed book sales could be falling for an entirely different reason. Nothing in the passage's reasoning necessarily means that e-books have attracted people *away* from hard copies.

Answers B) and D) are irrelevant because the issue at hand is whether e-books have attracted readers away from hard copies, not whether this is good or its economic impact.

Answers A) and E) are not flaws because for a critical thinking assessment we accept the reasoning given in the passage as true – assessing the quality of evidence is not part of the task, we are simply looking to see whether the reasons, if true, cause us to accept the conclusion.

## Question 4: B

The passage discusses how Governor Schwarzenegger is concerned with reducing expenditure, and how he proposes the use of the electronic devices as a means to cut the cost of school textbooks. The passage says that this is common sense, implying agreement with the governor. It is immediately apparent that this conclusion is *not* valid if the handheld devices do not save any money compared to textbooks, but nowhere in the passage is it stated that this is the case. Thus, B) is a valid assumption.

All the other answers are irrelevant because they do not directly affect whether the governor's suggestion to save money on textbooks is sensible. Thus, none of them are *required* to be correct for the conclusion to be valid, and thus they are not assumptions. In addition, answer D) is actually stated as a reason in the passage, and thus cannot be an assumption even if it was required for the conclusion to hold.

## Question 5: E

The passage describes experiments which have implied animals (pigeons) are better at maths than humans. However, it criticises these implications as "mistaken" and says that instead, the animals are simply learning from experience, whilst the humans are not. However, if we treat learning from experience as a mathematical process, then this criticism is not valid, and the experimental conclusions seem to be correct. Nowhere in the passage is it stated that learning from experience is *not* a mathematical process, so E) is a valid assumption.

Answers A) and C) are both over-concluding. The passage simply says how in this instance, the pigeons were learning from experience, and that therefore the conclusions that they are better at maths does not follow from the evidence. It does not say anything about whether the people were better or worse at calculating probability, so C) is incorrect. Equally, it doesn't make any claims about the ability to learn from experience (the fact the pigeons were doing it in this instance *doesn't* necessarily mean they are better in general at it) so A) is incorrect.

Answers B) and D) are not assumptions because neither is *required* to be true for us to accept the passage's conclusion. The passage is simply claiming that the experiments' conclusions do not follow from their results, because the process observed was learning from experience not mathematics. This is still a valid point even if B) and D) are not true, so neither of these are conclusions.

## Question 6: D

In order to calculate the *longest* possible journey time, we need to know is whether it is possible for the cyclist to have to wait at both sets of traffic lights:
- The question tells us that the cyclist cycles at 5 metres per second
- There are 900m between the 2 sets of traffic lights
- Thus, it will take the cyclist 180 seconds to move between the 2 traffic lights

We know that the two traffic lights are green for 120 seconds, both at the same time. Thus if the cyclist sets off from the first lights as they go green, they will arrive at the second set 60 seconds after they turned to red. Thus, they will have to wait 60 seconds at the second set until they turn green.

Thus, the longest possible journey is as follows:
- The cyclist takes 80 seconds to get from home to the first traffic lights, arriving exactly as they turn red.
- They then wait 2 minutes at the first traffic lights (120 seconds)
- The cyclist then takes 180 seconds to get to the second set of traffic lights
- The cyclist then waits 60 seconds at the second set of lights
- The cyclist then takes 100 seconds to get to the office.

Adding this up, we get a maximum journey time of 540 seconds, which is 9 minutes.

## Question 7: C

The simplest way to find the answer to this question is simply to calculate the price for each item in turn and add them up:
- ➢ We can clearly see that the dimensions for the first item exceed the maximum dimensions for a letter, so it will have to be sent as a parcel
  - The first 30 grams costs $1.22
  - The total weight is 300 g, leaving 270 additional grams to be paid for after the first 30 grams
  - 270 is 9x30, so we will need to pay 9x$0.17 (the price for each additional 30g is $0.17)
  - Thus, the total price for this item is $1.53+$1.22, which is $2.75
- ➢ The second item fits within the dimensions for the letter, and also weighs less than the maximum weight, so the cheapest way to sent it will be as a letter.
  - The first 30g costs $0.44
  - The total weight is 110g, so there are 80 additional grams to be paid for after the first 30 g
  - Each additional 30g or part thereof costs an additional $0.17. 80 is 2.67x30, so we will need to pay for 3x$0.17, which is $0.51
  - Thus, the total cost for this item is $0.95

Adding these costs together, the total cost for sending both items is $3.70. Thus, the answer is C)

**Note:** We are not given any information on what constitutes a postcard, and neither of the items are stated to be a postcard, so we can safely ignore the postcard information given in the table.

Also, perhaps the easiest trap to fall into in this question is attempting to calculate the proportion of postage for sending 2/3 of 30g with the second item. Reading the question clearly, we see that the $0.17 charge applies for each 30g *or part thereof*, so we simply add a full $0.17 for the 20g leftover for this item of post.

## Question 8: B

From the viewing angle given in the picture:
Options A) and E) are both the bottom side, in different directions.
Option C) is the top side.
Option D) is the right hand side, where left to right in C) corresponds with top to bottom from the viewing angle in the picture
Option B) does not correspond to any of the sides, so this is the answer.

## Question 9: C

The passage argues that steps must be taken to prevent child labour, and that nothing short of a ban is acceptable. One of its reasons is that children are more productive than adults, making child labour attractive to employers, so B) would *strengthen* the argument, not weaken it. Option E) would also strengthen the argument, suggesting that international pressure has failed to bring an end to child labour, so a ban may be necessary. D) also somewhat strengthens the argument by indirectly suggesting that child labour reduces the price of products, which is cited as a reason why child labour is attractive to employers.

A) does not strengthen or weaken the argument, and is actually an assumption in the passage, being required to be true for the arguments conclusion to be logical.

C) does weaken the passage as the passage discusses how child labour reduces the potential for education. However, if we accept C) to be true, then these children would not accept education without the wages from their work. Thus, C) is the answer.

## Question 10: E

The passage says that canal usage declined due to other, faster/cheaper forms of transport becoming available. It then says that the use of lorries is becoming less practical and more expensive, and that this means canals will soon be used again. However, if there are other forms of transport available as alternatives to lorries, then the decline of lorries will not necessarily lead to the resurgence of canals. Nowhere is it stated that there are no other alternatives, so E) is an assumption from this passage, and is the main flaw.

Answers A) through D) are discussing reasons why the road network may still be preferable to the canals. However, the passage is discussing why the decrease in road networks will result in a resurgence of the canals, so none of these answers affect the argument as strongly as E). Thus, E) is the answer.

## Question 11: A

The passage discusses how mentioning right and wrong in a statement does not alter the factual content of that statement, so A) expresses the main conclusion of this passage. Answer B) is referring to an example discussed in the passage to help explain/illustrate this point, and is not a main conclusion. C) and D) both disagree with the passage, which says claims about things being right and wrong are evincing moral approval/disapproval. Thus, neither of these are conclusions from the passage. E) is an intermediate conclusion, which then goes on to support the main conclusion given in A). We can see that if we accept E) as true, it gives us good cause to believe A), but this does not apply the other way round.

Hence, A) is the main conclusion.

## Question 12: B

None of the answers mention Barcelona as a possible destination, so we can instantly discount this row.

Answering this question is now simply a matter of proceeding through the other answers, calculating the price of each one, and subtracting this from £2000, to see how much is left over for spending money:
- Option A) would be £500 for the Ferry and train (£250 per person for 2 people), then £1200 for Hotel (£120 per night for 10 nights). This is a total of £1700, leaving £300.
- Option B) would cost £390 for flights, plus £1020 for hotel (£42.50 per person per night, which is £85 per night, for 12 nights). This totals £1410, leaving £590.
- Option C) would cost £480 for car hire and fuel (£40 per day for 12 days), plus £960 for hotel (£80 per night for 12 nights). This totals £1440, leaving £560.
- Option D) would cost £220 for flights, plus £1440 for hotel. This totals £1660, leaving £340.
- Option E) would cost £160 for flights, plus £1260 for the hotel. This totals £1420, leaving £580.

Of these options, B) leaves the closest to £600, so B) is the answer.

## Question 13: A

The volunteer wants exactly 5000 calories in the mix. Thus, the first step is to calculate how many calories are already present, to know how many calories are required from the sunflower seeds:
- 150g of mealworms, which are 150 calories per 100g, will provide 225 calories
- 150g of apples, which are 350 calories per 100g, will provide 525 calories
- 250g of raisins, which are 300 calories per 100g, will provide 750 calories
- 125g of suet, which is 800 calories per 100g, will provide 1000 calories
- Thus, the total number of calories already present is 1000+525+750+225 = 2500 calories.
- 5000 calories are required, so the sunflower seeds must provide 2500 calories.
- Sunflower seeds are 500 calories per 100g. 2500 is 5 times 500, so we need 5x100g of sunflower seeds, which is 500g.

## Question 14: B

As Freya runs to collect the stick, the distance between the 2 will increase. The distance will then decrease as Freya returns the stick to Sue. Thus, the graph showing the distance must go up then down to reflect this. Only the graphs in B) and D) show this, so the answer must be one of these 2.

To decide whether B) or D) is correct, we need to look at the rate at which the distances will change. During the collection of the stick, Freya is running away, whilst Sue is walking slowly towards Freya. Whilst the stick is being returned, Freya is running towards Sue, whilst Sue is still walking towards Freya.

Thus, we expect the distance to change faster whilst the stick is being returned than whilst Freya is collecting the stick, because during this stage the two are walking towards each other. Thus, we expect the gradient of the graph to be steeper whilst the distance is decreasing, during the returning of the stick. Graph B) correlates with this, whilst Graph D) shows the opposite. Thus, the answer must be B)

## Question 15: C

The passage describes how many people find the BBC license fee unfair, and describes 3 possibilities for continued funding of the BBC (continuation of the license fee, commercial funding or general taxation). It then goes on to highlight a problem with two of those possibilities (general taxation and commercial funding). The passage does not discuss how efficient/good at producing revenue the 3 possibilities would be, so D) and E) are not valid conclusions from the passage.
B) is also incorrect. The passage says how the commercial funding and general taxation funding options would put the BBC's independence at risk, but does not mention that this is the case for continuation of the license fee.

Option A) is not correct because the passage does not say any other options can be used for funding the BBC, it simply mentions how they would put the BBC's independence at risk. Thus, A) is not a valid conclusion.

Option C) is a valid conclusion. The passage mentions how many people do not like the licence fee, then criticises the alternatives. If we accept all these reasons as true, we have good cause to believe Option C). Thus, C) is the answer.

## Question 16: C

Option A) actually disagrees with the passage, which says that teeth left in bones *are* evidence for predation, so this is not an assumption. Option B) is also not an assumption because the passage states that the speed of a dinosaur is hard to assess, so we assume this to be true. Thus, B) is irrelevant. Option D) is a flaw in the passage, but it is not an assumption because the passage discusses several pieces of evidence why T-Rex may not be a predator. We do not need all of this evidence to be true for some of them to be valid, so D) is not *required* to be true for the passage's conclusion to hold. Thus, D) is not an *assumption*. E) is not relevant to the arguments reasoning, so is not an assumption.

The passage discusses evidence given by a palaeontologist that T-Rex was not a predator, and says that instead, T-Rex *must* have been a scavenger. However, nowhere is it stated that T-Rex could not have been both, and without this, the conclusion it must have been a scavenger is not valid from the reasons given. Thus, C) is an assumption from the passage.

## Question 17: C

The passage says usage of libraries has declined, but argues this is due to the style of libraries rather than a lack of interest in reading books. It concludes by saying this means the future of the current library functions will be better protected by restyling libraries.

Answer B) actually weakens this argument, as people looking for reference books are unlikely to be concerned with the style of the library. Answers E) also weakens the argument by suggesting that the rebranded library encourage use of *other* library functions, rather than preserving the current function of borrowing books. Answer D) is not relevant, because it does nothing to help determine whether the decline in library use is due to lack of interested in books or due to the poor branding of libraries.

Answers A) and C) both offer support to the argument, but Answer C) *most strengthens* the argument because it is discussing book usage in Britain. Answer A) discusses library usage in other countries, which may have patterns of library and book usage that differ from the UK. Answer C) refers to how the purchase of books in Britain has increased during the period in which library use has declined, offering direct support to the notion that poor branding of libraries may be responsible, rather than lack of interest.

## Question 18: D

The longest period of waiting for traffic on the side road will occur when they wait for the pedestrian crossing phase, then for the lengthened green period on the main road. During this period, the maximum length traffic on the side road can wait will be:
- A 2 second period after the side road lights going red, when all lights are red
- A 10 second period whilst the pedestrian lights are on green
- Another 2 second period after the pedestrian lights turn red. All lights are on red
- A 35 second period for the main road lights to be green
- A final 2 second period where all traffic lights are on red

This adds up to 51 seconds. Thus, D) is the answer.

## Question 19: D

We can see from the question that Ryan's Birthday (the earliest in the year) is on the $50^{th}$ day of the year, and that since it occurs before the $28^{th}$ February, it is unaffected by leap years.

All the other grandchildren's birthdays are affected by leap years, as they happen after the $28^{th}$ February. Thus, in a leap year the other birthdays will occur on the $124^{th}$, $151^{st}$, $251^{st}$, $322^{nd}$ and $351^{st}$ days of the year.

For another birthday to happen on the same day of the week as Ryan's, it must happen a number of days after Ryan's birthday that is a multiple of 7. Thus, we simply calculate how many days after Ryan's birthday each of the other birthdays occurs on, and see which one is a multiple of 7:
- 124-50 = 74. This is not a multiple of 7
- 151-50 = 101. This is not a multiple of 7
- 251-50 = 201. This is not a multiple of 7
- 322-50 = 272. This is not a multiple of 7
- 351-50 = 301. This is 43 times 7

Thus we can see that Robert's birthday happens 301 days after Ryan's, which is 43 times 7. Thus, Robert's birthday must happen on the same day. Hence D) is the answer.

### SECTION ONE — 2014

**Question 20: E**

The key to answering this question lies in looking at *how many times* each symbol in the code appears in the word, and where. Although the code changes each time Alistair writes, we can presume that within each message a given symbol always represents the same letter of the alphabet.

Looking at the symbols representing "Alistair" confirm this is the case. The 1st and 6th symbols are the same, just as the 1st and 6th letters in "Alistair" are both "a". Equally, the 3rd and 7th symbols are the same, just as the 3rd and 7th letters are both "i".

Now, if we look at the code for when Alistair is coming, we see two useful clues:
- The 3rd and 7th symbols are the same.
- The 4th and 8th symbols are the same.

Thus, we know that in the word representing when Alistair is coming, the 3rd and 7th letter must be the same, and the 4th and 8th letter must also be the same. The only word which fulfils this criteria amongst the options is "SOMETIME".
Hence, E) is the answer.

**Question 21: B**

The passage argues that university fees of £27,000 for a 3 year degree are reasonable, using a comparison with a similarly-priced car as justification for this point of view. Answer A) actually reinforces this point, suggesting that the degree is worth more, and therefore the fees are better value. Answer C) is simply a personal opinion that disagrees with the passage, and does nothing to affect the validity of its argument.

Answer D) is not a flaw because the passage states that the fees *will* lead to an increase in choice for students, so for the purposes of a critical thinking assessment we must accept this point as true. Answer E) is not relevant because the passage is discussing whether the *current* fee increases are reasonable. Future fee increases are irrelevant.

In contrast, answer B) correctly identifies that a person's car and a person's education are not intrinsically linked in anyway. Thus, discussing car prices when talking about education is irrelevant, and this removes a significant amount of the justification given in the passage for saying the fees are reasonable. Hence, B) is a flaw in the passage.

## SECTION ONE 2014

**Question 22: D**
The passage reasons that cats are interesting to most people in short doses, even if you are not a particularly keen cat lover. However, over long periods, cats are of interest only to a few people.

Answer A) claims that many subjects are interesting in short doses, but not over long periods. This is incorrectly extrapolating from the passage, which makes no claims about how many subjects are only interesting in small doses.

Answer B) follows the opposite pattern of reasoning, in which the first few bits (equivalent to the short film) are boring, but the further detail (equivalent to the longer film) is more interesting.

Answer C) reasons that many people may *not* share an interest in cats and/or babies. This is not the same as the passage, which claims that most people *do* have a short-term interest in these things, just not a long-term one.

In Answer D), the person is interested in short doses of photographs, but not in a long session of watching photographs. This is the same reasoning as the passage, where people may be interested in the short film but not in the longer one.
Answer E) describes whether cats are interesting enough to make a long film. This is not the same as the passage, which refers to the interest being dependent on the person watching it.
Hence, the answer is D)

**Question 23: C**
The issue raised with the University minister's comments is that a more qualified student is being charged more than a less qualified student. Thus, the less qualified student, by virtue of being less qualified, is getting a better deal. This principle is related to whether the students are being treated in accordance with their ability. Answer C) illustrates this principle, so this is the answer.

Answers A) and B) both discuss issues of timing, with no reference made to whether the later buyer/applicant is more deserving of the job/holiday, so these answers do not illustrate the principle used in the passage.

Answer D) is incorrect as the passage is actually discussing selling of education at different prices, not saying that it should not be sold. Answer E) is completely irrelevant, discussing logistical problems, and whether they should compromise the law.

# SECTION ONE 2014

**Question 24: D**

The fastest way to calculate the answer for this question is simply to calculate the largest expenditure possible, and then see what we can take off to get the amount as close to £28.50 without going over it.

The largest expenditure possible is as follows:
- Duck confit Salad for £6.95
- Rainbow Trout for £16.10
- Profiteroles for £6.20

This gives a total expenditure of £29.25. This is 75p over the allowance of £28.50. Thus, we are looking for the smallest reduction in price that is larger than 75p.

The smallest reduction in price possible that is larger than 75p is found by substituting Duck confit salad for smoked salmon, a reduction in price of 85p. This gives us a new total price of £28.40. Hence, D) is the answer.

**Question 25: C**

The first requirement of this question is to calculate how many calories Chris must burn in order to lose 7.5lb. The note at the bottom of the table says each lb in fat = 3500 calories, so to lose 7.5lb of fat, Chris must lose 26250 calories.

Now we need to see how many miles Chris is walking. He aims to walk 26 miles each week, for 10 weeks, giving a total of 260 miles of walking.

Now we can see how many calories Chris needs to lose for each mile of walking. Here, we can save ourselves some difficult calculating by taking into account a few useful points:
- 26250/260 is a difficult calculation to work out without a calculator. However, we can see that 260 multiplied by 101 is 26260, which is very close to 26250.
- Additionally, we can see that all of the values given in the table for calories burned per mile are integers.
- The question asks how many calories Chris needs to burn per mile to lose *at least* 7.5lb, *not exactly* 7.5lb.

Taking these points into account, we can round up the number to the nearest whole number, and assume that Chris needs to lose 101 calories per mile to lose the weight.

Finally, we look at the table and see which walking speed is required to lose 101 calories per mile of walking. We are told that Chris' starting weight is 180lb, so we simply look at that column in the table to find the slowest walking speed which will lose over 101 calories per mile of walking for a starting weight of 180lb.

We can see from the table that this walking speed is 4.0 mph. Hence, C) is the answer.

## Question 26: D

We can see from the first graph that Arthur has the most money at the start, and Carol has the least, and since they have all spent the same amount this order must also apply to the end of the year, so we can instantly discount Option C), which shows more money in Carol's account than Belinda's.

If we treat the amount originally in Belinda's as X, we can estimate the amount in Carol's account as somewhere between 0.5X and 0.75X. Equally, we can estimate the amount in Arthur's account to be somewhere between 1.5X and 2X. We cannot calculate precisely since no numbers are given.

We are told that half of Belinda's money has been spent, so we can treat the amount spent by each person as 0.5X. Thus, as the end of the year, the amounts in each account will be the following:
- Arthur's Account – Somewhere between X and 1.5X
- Belinda's account – 0.5X
- Carol's Account – somewhere between 0 and 0.25X

We can now see that the amount in Carol's account should be less than Half of the amount in Belinda's account, and the amount in Arthur's account should be between two and three times the amount in Belinda's account. The only option which fulfils this criteria is Option D), so this is the answer.

## Question 27: B

The passage describes using willow trees as fuel as *one of the most promising* solutions for high carbon emissions, and goes on to describe several advantages of using willow trees for fuel. It also mentions that this is already used in other countries, to further reinforce this point. All the reasoning given ultimately supports the idea that we should use willow trees to produce fuel. Hence, B) is the main conclusion of this passage.

Answers A), C) and D) are all reasons in the passage which support this conclusion, and are not themselves conclusions. Answer E) could be described as a conclusion, which follows on from the reasons in A), C) and D), but Answer E) itself then supports the Main conclusion in B). Thus, Answer E) is an *intermediate conclusion*, not the main conclusion.

## Question 28: B

Answer A) is not a valid conclusion. The fact that the police strike coincided with a doubling of the homicide rate *does not* mean that the end of the strike will cause the rate to halve.

Answer C) is not a valid conclusion because the passage provides two conflicting view points, neither of which are stated as fact by the other, and which explicitly disagree with each other. Thus, we cannot *reliably* conclude that C) is true,
Answer D) is not a valid conclusion because the passage makes no reference to the police officers' opinions on the pay rise, and neither the government nor the police officers have claimed that this was the reason for ending the strike.

Answer E) is not a valid conclusion because no reference is made in the passage to why the government decided to offer the striking officers an amnesty from punishment, or to the government's opinions on policy pay.

Answer B) can be reliably concluded. We are told that the police officers accepted a 6.5% pay rise, so we know they will be on improved pay. We also know that the strike has ended before the carnival, so the carnival will be staffed by a full-strength police force. Whether the strike ended *because* of the carnival is irrelevant to this, the important point is that it *did* end before the carnival.
Thus, B) is a valid conclusion from the passage.

## Question 29: B

The passage discusses how discussion between students, parents and teachers contributes to students doing well in exams by revealing problems. It concludes from this that if a school provide opportunities for meetings between these 3, it will contribute to its students achievements. However, it is not necessarily the case that that providing opportunities for meetings will lead to discussion, and this is not necessarily the case, and nowhere does the passage state that this is the case. If this is not the case, the passage's conclusions do not necessarily follow from its reasons, so Answer B) correctly identifies an assumption in the passage.

Answers C) and E) both strengthen the passage's conclusion, if true. However, neither is required for the passages conclusion to be valid, so they are not assumptions. Neither C) or E) is an integral part of the argument.
Answer A) is not an assumption, because the presence of other factors does not mean that highlighting emotional problems is not important. Thus, the passage's conclusion is not *dependent* on A) being true. Thus, Answer A) is not an assumption.

Answer D) is not an assumption because the passage discusses how it is import for students *not to suppress* emotional problems, and how the discussion between parents, teachers and students simply brings them out into the open. Here, the student's academic performance does not rely upon the problems being *solved*, it simply relies on them being openly discussed.

Thus, Answer D) is not required for the passage's conclusion to be valid, so it is not an assumption.

**Question 30: D**
We are told the following key pieces of information:
- Average speed of first Journey = 30 km/h
- Total time of first journey = 30 minutes (or 0.5 hours)
- Total mileage after 2$^{nd}$ journey = 24 km
- Average speed after 2$^{nd}$ journey = 32 km/h

This question is solved by using these facts and the equation of speed=distance/time as follows:

$$Total\ Speed\ after\ 2\ journeys = \frac{Total\ distance\ after\ 2\ journeys}{Total\ time\ after\ 2\ journeys}$$

*Let T = Total time after two journeys*

We know the Total speed and the total distance after 2 journeys, so we can plug these into this equation to calculate total time as follows: $32 = \frac{24}{T}$
This gives: $T = 0.75\ hours = 45\ minutes$
We know the first journey took 30 minutes exactly, so the 2$^{nd}$ journey must have taken 15 minutes.

Now we need to know the total *distance* of the 2$^{nd}$ journey. We know the total distance after both journeys was 24 km, and the first journey was 15 km, so the 2$^{nd}$ journey must have been 9 km.

Now we know that the 2$^{nd}$ journey was 9 km long and took 15 minutes (0.25 hours). Plugging these figures into the $Speed = \frac{Distance}{Time}$ equation, we get the following:
$Speed = \frac{9}{0.25} = 36\ km/h$
Thus, the answer is D)

# SECTION ONE 2014

**Question 31: B**
There are 10 possibilities for the first digit (0, 1, 2, 3, 4, 5, 6, 7, 8 & 9).
For each of these possibilities, there are 9 possibilities for the second digit which will not be the same as the first digit (e.g. for a first digit of 0, the $2^{nd}$ digit can be anything except 0).

For each of *these* possibilities, there are 8 possibilities for the third digit which will not be the same as either of the first two digits (e.g. if the first two digits are 0 and 1, the third digit can be 2, 3, 4, 5, 6, 7, 8 or 9).

The answer can be found by multiplying these 3 numbers: $10 x 9 x 8 = 720$.
Hence, the answer is B)

**Question 32: D**
This question is difficult to describe on paper, as it requires visualisation of 3-D shapes. The best way to prepare for these questions is to practice visualising nets, and then use cut-out nets to verify your answers (remember that you can always take scissors into your exam!).

**Question 33: A**
The passage discusses how people enjoy watching fictional violence on TV and in novels, but the recent increase in violent crime shows that increasing numbers of people are not satisfied with watching violence and need to take part in violence to be satisfied. However, Answer A) counters this by saying that most murders are not random killing of strangers to satisfy an urge, but pre-meditated and targeted at close family members. Thus, A) does weaken the passage.

Answer B) does not weaken the passage because the passage simply claims that increasing numbers of people are not satisfied with watching/reading about violence. It does not claim that an increase in violent films has caused the increase in violent crime.

Answers C) and D) are not relevant because the passage does not make any reference to the age of those carrying out crime, or whether the film classification system would make any difference.

Answer E) could actually be seen to strengthen the argument, because the argument refers to people not being satisfied with watching violence. If they are seeking real-world violence to instead satisfy their urges, then they would need to be able to tell the difference between real-world and TV.

# SECTION ONE  2014

**Question 34: C**

The passage describes a small group of anarchists causing violence during a presentation, but then goes on to discuss and criticise the action of describing all violent protestors as "anarchists".

Answer A) does not identify a flaw in this passage, because the passage does not claim that violent protestors cannot be anarchists – in fact it agrees with this. It simply disagrees that *all* violent protestors are anarchists.

Answer B) does not directly relate to the discussion about whether all violent protestors are anarchists, so is not a flaw in the passage.

Answer D) somewhat reinforces the passage. By questioning how possible it is to call someone an anarchist, it casts further doubt upon whether it is sensible to label all violent protestors as anarchists.

Answer E) also somewhat strengthens the passage, which claims that anarchists generally want an abolition of the state. If this group of violent protestors want an *increase* in state funding, then this suggests that they may not be anarchists.

Answer C) does detract from the passage, because the passage uses the fact that the violence occurred at a protest *against* decreased government spending as evidence that the violent protestors may not be anarchists (who the passage claims generally want an abolition of the state). However, Answer C) correctly identifies that the violent protestors may not agree with the protest's official aims, removing one of the reasons given why they may not be anarchists.

Hence, C) is the answer.

**Question 35: C**

The passage criticises government advice on milk, saying it makes false assumptions and ignores health benefits of full-fat milk. It then describes the advice as an example of unhelpful government intervention. This last sentence clearly follows from the reasoning given in the passage, so this is the main conclusion, and thus C) is the correct answer.

Answers A) and B) identify reasons given in the passage, which go on to support the conclusion given in C).

Answers D) and E) could be seen as conclusions from the passage, but both would require leaps of logic not made in the passage in order to be conclusions, and thus are somewhat criticisable as conclusions.

Also, if we accept either of these answers as true, they both go on to support the conclusion given in C), making them *intermediate* conclusions, not the main conclusion.

## SECTION ONE 2014

**Question 36: E**
First we calculate when she arrived according to her watch:
- The train was 10 minutes late
- The taxi was delayed 15 minutes due to the traffic
- Walking to the room took 5 minutes longer than expected

Thus, there was a total of 30 minutes delay. We know she intended to arrive exactly on time, so by her watch she will have been 30 minutes late.

We are told that the clock in the interview room was running 30 minutes slower than her watch, so according to this clock, she will have been exactly on time. Hence, the answer is E)

**Question 37: C**
First, we need to calculate what the cheapest option for membership will be under the conditions given in the price:
- Paying the discounted renewal price, because a change in membership counts as a renewal
- 12 visits during the year (Once a month for the full year)
- Locker hire (£2 a visit unless included in membership cost)
- Not entering any competitions

With 12 visits, the cost of a locker will be £24 throughout the year if not included in the membership price. Thus, the total cost of Bronze membership will be £32 (£8 for yearly membership plus £24 for locker hire through the year). The total cost of silver membership will be just the £28 renewal price, since a free locker is already included.

Thus, the cheapest option is a 1 year silver membership, costing £28.
We are told the person is currently a Gold member, that she made the payment 6 months ago, that she joined as a new member, and that the membership is about to expire. Thus, she must have paid the new member price for a 6 month Gold membership, which is £40.

Hence, the new payment will be £12 less than the payment 6 months ago, so the answer is C).

## SECTION ONE — 2014

**Question 38: A**

104 x 11-year-olds participated in swimming, compared to 150 16-year olds. This is roughly a 2:3 ratio of 11-year olds:16-year olds. Thus we are looking for another sport with a roughly 2:3 ratio of 11-year olds to 16-year olds. Starting from the top of the column, we see that the first sport mentioned, football, had a 120:181 ratio, which is also very close to a 2:3 ratio.

We can verify this by checking the rest of the ratios in the other sports:
- Cricket is 120:133, which is closer to 1:1 than 2:3
- Hockey is 55:66, which is a 5:6 ratio, also not close to 2:3
- Tennis is 123:149, which is roughly 5:6, and not close to 2:3
- Squash is 51:97, which is closer to 1:2 than 2:3

**Question 39: A**

The passage discusses how some commentators have cast doubt on the idea that government campaigns have led to a decrease in car accidents, claiming that instead many may simply not be reported. It uses the increase in hospital admissions as evidence to back up this claim.

If we accept all these reasons as true, we have good cause to believe that those thinking road safety is increasing continually may be incorrect, and thus A) is a valid conclusion from the passage. However, all the points refer to how it *may* be the case that road accidents have increased. Nothing is stated for certain, so we cannot conclude that the government initiatives *have* been unsuccessful. Thus, B) isn't a valid conclusion.

Answer C) could be described as an assumption in the passage. The main supporting evidence in the passage is that hospital admissions have increased, and if this is true the argument's conclusion no longer securely follows from its reasoning. Thus, C) is an assumption, and not a conclusion. Answers D) and E) both *counter* the passage's argument, and are thus not conclusions.

# SECTION ONE — 2014

**Question 40: C**
The passage argues that the UK government's policy of increasing university places is appropriate, arguing that graduate unemployment has decreased and the UK is wisely responding to the demands of a modern "High-Tech" economy.
Looking at the reasoning used, we can see that just because the graduates are getting jobs does **not** mean they are getting jobs *requiring degrees*. If this is the case, then the passage's reasoning no longer supports the conclusion that the extra university places are required for the economy's demands. Thus, C) is a valid assumption of the passage.

Answers A) and E) both refer to whether increasing university places is always a requirement or a response to economic growth. This does not directly affect the conclusion regarding whether the UK's increase is required to be a high-tech economy. Thus, neither of these are assumptions of the passage.
Answer B), if true, would reinforce the conclusion, providing additional strength to its reasoning. However, it is not *required* to be true for the conclusion to be valid, so this is not an *assumption* of the passage.
Answer D) would actually *weaken* the passage if it is true. Thus, it is not an assumption of the passage.
Hence, the answer is C)

**Question 41: B**
The passage discusses how more and more crimes are being committed by girls and how for several crimes, the increase is seen as linked with an increase in alcohol consumption, concluding that this means greater alcohol is responsible for the increase.
Answers D) and E) would both *strengthen* the passage. D) reinforces the point that girls are committing crimes not traditionally associated with them, whilst E) describes an increase in alcohol consumption. If these are both true, the passage's conclusion is reinforced.

Answers A) and C) are irrelevant. The passage does not refer to whether girls are committing more and more crime relative to other age/gender groups, it simply states that they are committing more crimes. Thus, the effects of alcohol on other groups are irrelevant to the passage's conclusion.
Answer B) does weaken the passage, as it suggests that the chances of prosecution have increased for girls. This provides another reason why the number of recorded crimes committed by girls may be rising, weakening the passage's conclusion that alcohol is responsible.
Hence, the answer is B)

# SECTION ONE         2014

**Question 42: C**

This question asks us to find the value for y, the 5-year moving average for the year of 1909.

We do not know the average maximum day temp for 2009 itself (value x), or next year's 5-year moving average (value z)

However, we *can* calculate the value of x, using the 5-year moving average for 1913. This average will be composed of the average of the mean max temps for the Years 1909,1910,1911,1912 and 1913. Since the table provides the information for the mean max temp for 4 of these years, we can use this information to calculate x as follows:

- 5 year moving average for 1913 $= \dfrac{x + 13.3 + 14.9 + 13.7 + 14.2}{5}$
- $13.82 = \dfrac{x + 13.3 + 14.9 + 13.7 + 14.2}{5}$
- $69.1 = x + 13.3 + 14.9 + 13.7 + 14.2$
- $69.1 = x + 56.1$
- $69.1 - 56.1 = x$
- $13.0 = x$

Now, we can use this value, along with the values of the years 1905-1908, to calculate the value of y, following a similar procedure to before:
- $y = \dfrac{13.4 + 14.4 + 13.4 + 13.7 + 13.0}{5}$
- $y = \dfrac{67.9}{5}$
- $y = 13.58$

Hence, the answer is C)

**Question 43: C**

To answer this question, we must calculate the average number of fish per angler from each river *during each year* – since the question asks what the highest figure is *in any one year*, we cannot average these figures out over both years.

River Dark:
- Year 1 = 77 scale sets from 5 anglers = 15.4 per angler
- Year 2 = 125 scale sets from 30 anglers = ~4 per angler

River Fare:
- Year 1 = 85 from 20 anglers = 4.25 per angler
- Year 2 = 35 from 8 anglers = ~4 per angler

River Gwynt:
- Year 1 = 71 from 19 anglers = >4 per angler
- Year 2 = 132 from 2 anglers = 66.5 per angler

River Tine:
- Year 1 = 105 from 43 = ~2.5 per angler
- Year 2 = 66 from 19 = ~3.33 per angler

River Yarrow:
- Year 1 = 80 from 18 = >4 per angler
- Year 2 = 150 from 25 = 7 per angler

*Notice that these calculations have frequently made use of estimation for tricky calculations here, which could be time consuming to work out exactly. We do not need to work them out exactly as the question only asks which individual figure is highest.

For example, once we have seen that in Year 1 River Dark returned 15.4 scale sets per angler, we can clearly see that the figure for the 2$^{nd}$ year was lower (~4). We do not need to work out the exact answer (4.1666) to see that 125/30 is less than 15.4, so we can save time by simply estimating approximate answers.

We can clearly see that the highest individual figure is year 2 for River Gwynt. Hence, C) is the answer.

# SECTION ONE — 2014

## Question 44: A
We are told that **two** of the lights in the display are not working, so we are looking for a number which, in this matrix, could be turned into an "8" by the addition of **one or two** extra lights being illuminated. We are told that the lift has **left** the ground floor, so he **cannot** be on the ground floor.

We can see that in the matrix show, figures 2, 3, 5, 6 and 9 could be turned into a figure 8 with 2 more lights being illuminated. Thus, there are 5 floors he could have got out on by mistake.

Figure 1 would require 5 extra lights, Figure 7 would require 4 extra lights and Figure 4 would require 3 extra lights to be turned into Figure 8, so none of these are possible. Hence, there are 5 possible floors and the answer is A)

## Question 45: C
The passage argues against the claim that cooked food is less healthy due to loss of vitamins, claiming we should be "thankful" our ancestors developed cooking. If we accept all the reasons given in the passage to support this claim, we have good cause to believe this claim. Thus, C) expresses the main conclusion of the passage.

E) is also a valid conclusion from the passage, but it goes on to support the statement in C). Thus, E) is an intermediate conclusion in this passage, *not* the main conclusion. A) could also be seen as an intermediate conclusion, supporting the main conclusion given in C).

Answers B) and D) are reasons given in the passage to support its conclusion.

## Question 46: B
The reasoning in this passage can be described as "A is required for B. A is happening, therefore B will be possible". Here, "A" is the light level being low, and "B" is stargazing being possible.

- A) could be described as "A must happen for B to happen. A isn't happening, so B won't happen", with A being the train being on time and B making the flight. This is not the same as in the passage.
- B) reasons as "A must happen for B to happen. A is happening, so B will happen". This is the same as in the passage, so B) is the answer.
- C) reasons as "A is required for B to happen. A has not happened, so B must not have happened either", where "A" is the exceptionally high tide, and "B" is a flood happening. This is not the same as in the passage.
- D) reasons as "A must happen for B to happen. A didn't happen, so C couldn't happen", where "A" is the water being boiling, "B" is making an excellent cup of tea, and "C" is the aunt enjoying the cup of tea.
- E) reasons as "the best produce comes from France, therefore produce from France must be the best". This is not the same as in the passage.

# SECTION ONE 2014

**Question 47: D**

The passage discusses how in boxing, the negative events (i.e. damage to the players) are the object of the sport, rather than an unfortunate effect of something going wrong. It claims that because of this, boxing is unacceptable whilst other sports are not. Thus, the principle underlying the passage is that 2 things are not the same if one is designed to cause negative events, and one has risks of negative consequences if something goes wrong.

Answer D) follows a similar principle, in which negative things only occur from bonfires when something goes wrong. Thus, it should not be banned in the same was as arson, where negative things are the objective.

None of the other answers refer to differences between intentional negative consequences and problems resulting in unintended negative consequences, so these are not the principle in the passage.

**Question 48: C**

The average number of great-great-grandparents would be 16 (2 for each great-grandparent).

However, we are told that 2 sisters from one family married 2 brothers from another among her great-grandparents.

The two sisters would have had the same 2 parents. We would normally expect the 2 great-grandparents to account for 4 great-great-grandparents, so we subtract 2 from the average total. The same applies for the 2 brothers, with another 2 subtracted from the average total here.

Thus, we subtract 4 from 16, leaving us with a total of 12 great-great-grandparents. Hence, the answer is C)

## Question 49: B

The maximum income from students will come when all canoes are being used by students. We are told that the school has 18 canoes, with the instructors expected to use their own. Thus, the maximum income will be from a group of 18 students. For a two hour session, each student will pay £12 (£7 for the first hour and £5 for the second hour). For 18 students this will be a total of £216 of income from the students.

The minimum instructor : student ratio is 1:6, so a group of 18 students will need 3 instructors. The instructors are being paid £6 an hour, so this will be a total of £36 paid to the instructors. £216 - £36 = £180 of profit for the school for this session.

## Question 50: B

We can see from the bar chart that the three groups accounted for approximately the following amounts of revenue:
- Adults = Just under £400
- Children = Just under 150
- Senior Citizens = Just under £150

Thus, we expect that the following numbers visited the Museum:
- Adults = ~400/7 = ~60
- Children = ~150/4 = ~40
- Senior Citizens = ~150/5 = ~30

Thus, we are looking for a graph which shows the number of adults, children and senior citizens visiting as a roughly 6:4:3 ratio.

Thus, it should clearly show the following things:
- Adults should account for < 50% of visitors (so A, D & E are incorrect)
- The number of senior citizens should be smaller than the number of children (so Graph C) is incorrect

The only graph left which fulfils these criteria is B). Hence B) is the answer.

**END OF SECTION**

# Section 2

*How should we evaluate advances in science?*

This is a difficult question in that it requires an in depth knowledge of both philosophy and science, and the interaction between them. It can be made simpler by clarifying in your opening paragraph the manner in which you are going to make the argument – i.e. from the points of view of profit and of societal benefit. Never the less, it remains a challenging question.

## *Introduction*
- There are two main ways to evaluate advances in science – either through the amount of people the scientific advance helps in regard to lives saved or some other metric, or through the amount of profit that it creates for the scientist/company who invented it.
- From a moral point of view, the importance of a scientific advance is clearly the former – how it helps improve people's lives.
- From an economic, and possibly a more realistic point of view however, the reality might be different. In our capitalist society, profit is a huge driver of scientific advances. Whether this is a good thing or not is a good way of addressing wider political issues while answering this question.
- A key issue at the heart of this question concerns what the purpose of science is – should it be to improve lives or simply to make profit?

## *Assessing profit in science*
- Due to the involvement of private, profit making corporations in the scientific field, you could argue that this is no longer the most important. A firm which creates scientific advances which make no money will quickly go bust and stop producing advances. This is an issue that you could explore morally – should this be the case? What are the implications for society? What are the implications for the future of scientific research?
- You could also explore the implications for having research publicly funded versus research which is privately funded. You could argue that private research funding is more likely to flow to issues which generate the most money rather than what is best for the people, leading to advances in areas such as arms and weaponry which might be at the detriment to society as a whole.

## Counter argument

The other side to this argument is that you could argue this is a false dichotomy, and that given the nature of scientific research industry, any invention that helps people sufficiently will make money anyway. Should this be the case? Try and assess the issue from a moral rather than a purely scientific/profit-based viewpoint.

## Time lag

Another angle of assessing the profit-based view of science is through timeframe – an advance might not be immediately profitable or beneficial, and it might take many years before an advance is recognised as being useful. An argument therefore could be made that focuses more on an advancement's long term impact rather than immediate impact.

## Conclusion

➢ Make sure that you include potential counter arguments for any of the above points that you make – it is vital that your essay comes across as balanced and not one-sided.

➢ In your conclusion you could argue that this distinction is not absolute – for example a scientific advance could both save lives and make profit for a company.

➢ Another distinction you could make in your conclusion is between how we ought to evaluate advances, and how they are actually evaluated. How could the current situation in regard to the profit motive be improved? Could, for example the government introduce legislation to prevent the negative effects described above? If so, what kind of laws would they be? Exploring these political implications is a good way to end the question.

## SECTION TWO     2014

*How much should we test a medicine before making it widely available?*

When answering this question, you should keep in mind the ambiguity of the question. Is 'we' whoever is producing the medicine, the government, or an independent medical authority of some kind? This will affect what kind of tests are had in mind. If it's not widely available before extensive testing, does this mean it is limited to certain people before this point, and they are the test? In this case, the access to the medicine is limited because those who take it will require very close monitoring. Is it unfair to limit access to a potentially life-saving medicine in this way?

Before beginning to answer, consider these questions; because of the time limit, you will most likely have to choose one or two such scenarios to discuss. You should choose the ones that you think are most ethically interesting – perhaps the borderline cases. For example, you might argue that we should test a medicine as much as possible to ensure that it is safe before making is available to anyone, but untested medicine should be available to people who are dying, as a last shot chance at life, as there are no risks involved for them, but it has the potential to help them live and to further medical science.

### Introduction:

- Begin by defining your terms, as above. Indicate what you consider to be 'widely' available, and whether tests in this instance would include human tests (i.e.: are you imagining untested medicines to be available to a limited number of people?)
- Clearly state your opinion within these defined parameters, and outline the main arguments you'll be discussing.
- Though it may seem obvious, it is worth briefly discussing *why* we test medicine before making it available to the public, and why your opinion either finds the balance between flooding the market with untested drugs and over-cautious withholding, or why the extremes of a complete lack of testing, or extremely rigorous testing, are in fact desirable.

*Potential arguments in favour of less testing:*

- You could make an argument for bodily autonomy and informed consent; so long as a patient know that a medicine is untested, and knows the risks of taking a medicine about which we know very little, they should be able to make the informed decision to try it.
- Many medicines treat fatal conditions; if a patient is terminal anyway, they ought to be given the opportunity to take a potentially life-saving drug.
- The only testing that really holds weight is human testing – testing on animals is flawed as their biology is so different to ours. As above, when a drug is deemed safe enough to test on humans, participation in these tests ought to be *widely* available, as limiting a potentially life-saving treatment to only a few people for doctors to experiment on is unfair.
    - **Counter-argument**: because of the arguments below (patient safety, doctors need to monitor patients on new drugs), it ought to be limited. However, as soon as human trials are successful, it ought to be widely available.
- Conversely, you could argue that human testing isn't nearly as important as people make it out to be. We can test on animals to measure the safety of a drug with great success, and we can now also test on artificially grown human organs, the results of which are strongly indicative of how a medicine will actually react in a human body. When these tests are successful, we ought to treat a medicine as being as thoroughly tested as if it had actually been in a human body, and make it widely available.

*Potential arguments in favour of more testing:*

- Patient safety ought to be a priority. While a new drug that is entering the human testing stage might be an amazing treatment in a few years, it's unlikely to be a ready, finished product during the initial testing stages. Limiting availability to a few test subjects also limits the damage that a new drug can do before it is refined after the results of human trials.
- On this note – patients subject to relatively untested drugs need to be closely monitored, precisely because the medicine is new and untested. It is simply not feasible for doctors to give that kind of attention to the huge number of patients that might request access to a new medicine for a common condition.
    - Counter-argument: isn't this an argument in favour of more funding for health services and training medical professionals so that more monitoring of patients that need it can be provided, rather than an argument against providing necessary medicine?

- Patients might think they're making an informed decision about their body when they choose to try a new medicine, but they're simply not as educated on medical matters as their doctors and the medicine's developers. The decision to let humans take a medicine ought to be up medical professionals.
  - **Counter-argument**: this is paternalistic. While doctors might understand the risks involved, they are able to convey these to a patient, and only the patient can decide whether these risks are worth it for them.
  - **Counter-counter argument**: the patient is naturally more emotionally invested in this, and might not be able to make a clear-headed decision, especially if it's a potentially life-threatening illness, and could grasp at straws.

- Further, many of the arguments in favour of less testing make it out to be a life or death situation every time, but this is melodramatic. Most medicines treat conditions that aren't fatal, and many new medicines treat conditions that are already treated in other ways. It would be fool-hardy to risk serious side-effects trying out a new drug on otherwise fairly well cared for patients before it's tested enough to be declared sufficiently safe. (You might here give your argument some depth by saying that terminal patients are an exceptional circumstance, in which you might favour the use of untested drugs).

## *Conclusion:*

- Restate your opinion, and summarise your arguments.

- Be sure to link your conclusion back to the question by referencing the purpose of testing before making medicine available, and why you feel the arguments you've favoured outweigh the arguments in favour of more/less testing.

- Given that you will most likely have had to restrict what you are saying to a few specific situations, it might be nice to make reference to other scenarios in which your opinion would be different. For example, if you've argued that making relatively untested medicine widely available is necessary as only human subjects can provide accurate information on the medicine's usefulness, you might add that, of course, if some other method were developed to test new medicines – computer simulations of chemicals affecting human bodies, for instance – then you would revise your position. This indicates that you've thought about scenarios you haven't been able to discuss in detail in this essay.

## SECTION TWO — 2014

*Should statistics be a compulsory subject of study at school? Why, or why not?*

In this sort of question, 'statistics' is, in part, a stand in for any subject – much of your essay should be spent arguing for a set of criteria for a compulsory subject of study, focusing on the criteria that statistics fits if you are arguing in favour of it being compulsory. Alternately, you could argue that no subject of study should be compulsory – in this case, you should make it clear why you believe this is the case and reject counterarguments, and use statistics to demonstrate your case (e.g.: you might argue that no subject is necessary for every day functioning, and then explain why statistics specifically isn't necessary). It is probably worth restricting your answer to secondary education, or even just the second half of secondary education.

Not only is statistics completely irrelevant to primary education (one is unlikely to prescribe making it compulsory for under 10s), but it is often also not engaged with in any depth until the later secondary years. You can mention, when answering the question more broadly, why a subject ought to be compulsory at any level (because it's necessary in later life, but also appropriate for that age group, etc.), but your answer ought to be focused on the areas to which the question directly applies – in this case, from the age when statistics generally becomes appropriate to study in depth and older.

### *Introduction:*
- Link your answer into a broader understanding of the concepts mentioned in the question by stating why a subject (if any) ought to be a compulsory subject of study at school, and thus whether statistics does not does not fit this criteria.

### *Potential arguments for:*
- Statistics arise in day to day life more often than we realise: it's used to back up scientific data that we read about and believe in the newspapers; it's used to predict the results certain actions will have; it's used to manipulate data so that people believe what the manipulator wants them to believe. A more complex understanding of statistics would help us make more informed choices in these circumstances.
  - **Counterargument**: art and music also arise in day to day life, and we could equally make the argument that a more complex understanding of these means of expression would help us understand each other better and lead us to have more fulfilled lives – as social and expressive animals, these are both hugely important – but one is far less likely to prescribe art as a compulsory subject of study.

- ○ **Counter-counterargument**: if art is truly as essential as statistics then the answer is to make the study of both compulsory, not neither.
- ➤ Studying statistics doesn't only benefit us where statistics can be directly applied. It involves logical thinking which helps develop our reasoning powers, which can be applied to any number of areas of study as well as day to day life.
- ➤ The kind of abstract logical thinking that it's involved in statistics isn't found in other areas of study in school. Geography, history, philosophy, politics, economics, etc. all involve the same sort of reasoning applied to different sets of knowledge, and the applied sciences (physics, chemistry, biology) involve very similar reasoning abilities. Statistics involves using very precise, unbiased logical reasoning to reach necessary conclusions – an education in this way of thinking is found by studying any other subject.

*Potential arguments against:*

- ➤ No subject of study should be compulsory past primary education. While it is true that basic reading, writing, and arithmetic are extremely helpful, if not necessary, for functioning in the world, these skills should be acquired by the time someone starts secondary school, or at the very least before they begin studying advanced enough maths that statistics is a subject of its own.
- ➤ The reason we generally don't prescribe art as a compulsory subject (see counterargument to the first argument in favour) is that we recognise people's differing abilities in varying fields. If we can see that some people have less interest in and predisposition towards artistic study, and so should not be forced to fruitlessly spend time on it, we ought to see that the same is true of statistics.
- ➤ Forcing people to study statistics because we feel it's in their best interests for the reasons described in the arguments in favour section is not only paternalistic, it also cuts into the time they have to study subjects that actually interest them, and that will benefit them specifically (i.e.: we can recognise that statistics is broadly a good thing to be studied, without forcing every person who wants to, say, be a historian, to study it, while taking time away from their study of history).

*Conclusion:*

Once again, engage with the question on a wider level by restating why any subject should or shouldn't be a compulsory subject of study, and describe how statistics fits into this.

## SECTION TWO — 2014

*Should Western nations refuse to trade with countries in which child labour is used?*

There is some ambiguity in this question, regarding what constitutes a 'Western' nation. Most likely, they mean rich economies – also known as the developed world, the first world, and MEDCs. It is worth noting briefly in the introduction exactly how you interpret 'Western world' to avoid confusion, and to show you have thought about it. It is also worth briefly defining 'child labour', since this probably doesn't mean any work by children under 18. A definition such as "legal full time work engaged in for the sake of sustenance by people under the age of 16" would suffice. Before answering, consider carefully the economic, social, and symbolic impact of such a refusal to trade, and your priorities regarding economic flourishing, human well-being, and international cooperation.

### *Introduction:*
- Clearly state your opinion and outline your arguments for it, as well as your reasons for dismissing counterarguments.
- Although not addressed in the question directly, you ought also to state whether you think child labour is a good, bad, or neutral thing. This will obviously affect whether you think we ought to refuse trade with countries in which child labour is used; it might be that you think we shouldn't because child labour is acceptable, or it might be that you think we shouldn't because, although child labour is unacceptable, a refusal to trade won't help. Addressing this issue will help give your argument more nuance if you're arguing for trade, or a more substantial moral backing if you're arguing against it.

### *Potential arguments for:*
- By participating in trade with nations in which child labour is legal, they not only implicitly endorse these practices, they also directly fund them.
- By refusing trade, they put pressure on Governments, who rely on healthy economic activity, to change their laws and end child labour.
- Refusing trade with nations in which child labour is legal won't only stop the funding of child labour, it will increase funding to more morally acceptable industries (since people won't stop demanding these goods, they will simply acquire them elsewhere), which will increase employment among adults.
- Western nations have higher standards of living and better working conditions; they ought to set an example for other nations. These nations realise that child labour is wrong, and have made it illegal for their own children to work; trading with nations in which child labour is legal is hypocritical, and shows a xenophobic prioritising of their own children above children in other countries.

> Western nations are among the strongest world economies; they have an unmatched ability to use this economic influence to make a real difference to economic activity in other countries.

*Potential arguments against:*
> Why do 'Western' nations have moral superiority regarding how our goods are produced? Though child labour is widely outlawed, workers often face dangerous conditions, long hours, earn lower than living wages, and aren't guaranteed sick leave.

> Refusing to trade with countries that use child labour won't actually raise the living standard for these children – it will put them out of work, and they will be left with even less.
> - **Counterargument**: without child labour to rely on, supply of labour in these countries will fall, and the price of labour (wages) will subsequently rise, as labour will be in higher demand relative to supply. Wages for adults will therefore be higher, which will benefit the working class of the country, and adults are more likely to be able to support their children.

> Punishing whole countries, often very poor countries, for industries that use child labour isn't fair; we ought to simply not fund goods that were produced by child labour.
> - **Counterargument**: the Governments that run these nations get tax money from every industry. By supporting any of them, we support the Government and its laws. By refusing trade, we threaten the Government's tax revenues and put pressure on them to change their laws.

> Child labour occurs in the majority of nations, and in particular in nations where manufacturing industry dominated the economy. Refusing to trade with all these nations simply isn't sustainable; we don't produce enough ourselves to fulfil demand.

> While children shouldn't be working, refusing to trade with nations where this happens will simply damage our relationship with them. The children will continue to work, and we will have no influence over it. We ought instead to negotiate with these countries to change their laws and use our economic pull to pressure them into doing so.

*Conclusion:*
Link your conclusion back to the question by referencing your definition of child labour, and discuss why children should or should not be working in such conditions, and why a refusal to trade with nations in which this happens would or would not prevent it, or whether other factors outweigh this.

# SECTION TWO  2014

*In our country, every citizen has one vote. A scheme is proposed which allows anybody to buy additional votes if they want to, with the proceeds being used to pay for good causes. Would this be a good scheme?*

The sort of question that asks you to discuss a hypothetical such as this is asking you discuss the issues surrounding the concepts it makes reference to. Here, those are the concepts of democracy and a good cause. Before answering, consider your feelings about both, the reasons they are widely considered beneficial (and whether you agree), and which, if either, you prioritise. Try to consider the full social, economic, and political impact such a scheme would, or could, have.

## Introduction:

- Clearly state your opinion and outline your arguments for it.
- Link your argument to the concepts of democracy and good causes as discussed above – what are the underlying principles and values that have lead you to hold the opinion that you do? Simply arguing that, for example, the scheme will corrupt democracy, is not enough; you must defend the implicit position that democracy is something that ought not to be corrupted.

## Potential arguments for:

- Many things that are widely considered 'good causes' – free health and education, food banks, care for those unable to care for themselves, etc. – are currently hugely underfunded.
    - **Counterargument:** the solution is not to sell off democracy but, rather, to acknowledge the wealth gap between those who rely on free services and those who have enough money to hypothetically pay for votes, and forcibly correct it.

- This scheme would motivate people to put their money towards these causes in exchange for something they really care about; political pull. Importantly, political pull doesn't cost money to make. Unlike, to use a simplistic example, a bake sale, where what is sold has a cost in materials and labour power, political power has no production cost. In the case of the bake sale, we are essentially relying on people to give more money than must be spent on the material costs of making the goods, which they do because they understand the money is going towards a good cause. This relies on altruism. The proposed scheme does not rely on altruism; in exchange for money, people acquire something they really want, and do not pay more than what is considered its worth. Since people are fundamentally selfish, this will be far more effective than other means of raising money.

- o **Counterargument:** but currently people DO put their money towards good causes without selfish motive, and without any personal acquisition as desirable as political power.

➢ People with money have the right to use it how they wish. They can buy power in other ways – hiring staff, donating to political causes, setting up activist groups – why do we baulk at the idea of selling votes?

➢ People with a lot of money deserve to buy votes; they're rich for a reason, because they worked hard and were clever enough to get rich. Since political activity affects economic activity, they have a right to use their well-earned wealth to affect how the economy will be run.

*Potential arguments against:*

➢ This scheme clearly undermines democracy by allowing the rich to buy more votes than the poor can afford, so they can influence Government unfairly.

➢ Since they influence Government, who are presumably who decides what constitutes a 'good' cause, they also influence where the money they use to pay for votes goes – it could easily be back into funds that benefit them. They would get richer, and subsequently be able to buy more political power, which in turn would make them richer.

➢ There is already a power imbalance between the rich and the poor, this scheme would only exacerbate it and make it worse for the reasons stated above.

➢ Currently, poorer people are still able to exert political pressure by pooling resources and taking collective action. Richer people can act independently, due to their individual affluence, but, despite being able to donate to campaigns and organise political action groups, etc., they don't hold the pull that large groups of people have; they can exert influence precisely because they are many in number, and their cause is widely supported. This scheme would undermine this dynamic.
  - o **Counterargument:** couldn't poorer people pool their affluence to buy more votes together?
  - o **Counter-counterargument**: even if they did, they wouldn't hold the pull that one rich individual did. This is partly because of the huge wealth disparity in our country, and party because the votes per person would still be much smaller than the votes assigned to one richer person. Each poor person would therefore still have far less political power than each rich person.

- ➤ This is key to political health – that a decision is made because it benefits the majority. Voting is the way in which we measure public preference. This scheme would defeat the entire object of voting at all.

*Conclusion:*
- ➤ Restate your opinion and summarise your arguments for it.

- ➤ Link your conclusion back to the question by making reference to democracy, and the concept of a 'good cause'.

- ➤ It's nice to conclude this sort of question, as discussed in the opening paragraph, with a sentiment regarding the concepts discussed as a whole, and the conclusions you had to reach to make the arguments that you did. For example, you might conclude that democracy is a universal good which is necessary to protect against corruption by the powerful. Or, you might conclude that charity and redistribution of wealth are far greater virtues, for the sake of which we can have some humility and sacrifice some political power.

**END OF PAPER**

# 2015

## Section 1

**Question 1: E**
The passage discusses reason why it is important to realise that drug crime is not solely related to illegal drugs such as heroin and cocaine, but also includes counterfeit and homeopathic medicines, citing examples of how each can also be harmful. It then suggests that previous efforts to tackle drug crime have not included counterfeit medicines, and this approach should be changed. If we accept the other reasons in the paragraph as true, we have good cause to believe this claim and so and E) must be the main conclusion of this passage.

Answer A) specifically focuses on the first line of the passage, and so does not encompass the real argument of the passage, which is to tackle counterfeit drugs.
Answers B) and D) are both reasons given in the passage, which contribute towards supporting E). Thus, they cannot be main conclusions.
Answer C) is irrelevant because homeopathic remedies are not the central issue of the text. Although the passage says that they do not work as advertised, there is nowhere in the passage which states that they are not effective, and so C) is false.

**Question 2: D**
The simplest way to do this question is by using algebra.
  ➢ Let x = number of 50p coins  and Let y = number of 10p coins
The information provided allows us to construct two separate equations which we can us together to find x and y.
We know that the total value of the coins added must be £8.50 and so:
  $50x + 10y = 850$. This can be simplified to $5x + y = 85$
Secondly we know that the total number of coins added is 25 and so:   $x + y = 25$
We can then solve these simultaneous equations by elimination:
  $4x = 60$   Therefore $x = 15$ and $y = 10$
This tells us that the woman inserted 15 (50p) coins and 10 (10p) coins into the machine. However the question asks us what combination is **retained** by the machine. As 5 (10p) coins are returned, the machine must retain all 15 (50p) coins and 5 (10p) coins.
Therefore the correct answer must be D, the machine retains three times as many 50s and 10s. Note that if you ignored the fact that the question is specifically asking for the combination retained by the machine, you would have instead gone

for answer B, thus highlighting the importance of reading the question very carefully.

**Question 3: B**
This passage discusses how in recent years voting has declined to such an extent that less than half of the population actually voted for the party currently in office, and so argues that the governing party does not have the support of the majority of people. B) correctly identifies this conclusion as not valid – it suggests that the party may still have the support of those who did not vote, and so would have an overall majority. Therefore B) represents the greatest flaw in the argument.

Answers C) and D) are irrelevant because the issue at hand is whether the current government has the support of the majority of the people, not why the voting has declined or who is responsible for the lack of legitimacy.

Answers A) and E) are not flaws because for a critical thinking assessment we accept the reasoning given in the passage as true – assessing the quality of evidence is not part of the task, we are simply looking to see whether the reasons, if true, cause us to accept the conclusion.

**Question 4: E**
The passage discusses how there appears to be double standards in the advertising industry which has allowed certain retail groups to utilise sentimental marketing techniques to outperform all of their rivals. The passage explains how John Lewis' advertisement has been extremely popular online, despite using themes that people may describe as sexist or out-of-touch, as it displays a more stereotypical view of a woman. Although there is evidence that A) might be true, this statement is far too bold to be supported and does not encompass the whole argument, which is about the use of sentimental marketing techniques.

There is no evidence to support B) and D) in the passage even though these might be underlying assumptions of the argument. In addition, C) cannot be right as it represents the stereotypical, politically incorrect view that women are trying to escape from. Hence E) in the main conclusion of the passage.

**Question 5: C**
This passage discusses how the rise of extremist parties across Europe in the 1920s and 1930s was a major factor in starting the Second World War. Furthermore, it warns that as extremist parties have recently gained more popularity, it is possible that history may repeat itself and lead to new conflicts in the near future. The passage then suggests that the only way to reduce the likelihood of future conflicts is to educate children about the dangers associated with extremist parties.

It is immediately apparent that the conclusion of the passage is not valid if educating children about the dangers of extremist politics does not cause fewer people to vote for them, but nowhere is it stated in the passage that this is the case. Thus C) must be the underlying assumption of the argument.

Answers A) and D) are both over-concluding and are too strong to be underlying assumptions. For example in the case of answer A), although the passage suggests that it is possible the dangerous situation will get worse, there is no concrete evidence which proves that history will definitely repeat itself. Furthermore, although extremist parties were directly involved in starting the Second World War, there is no evidence that the main reason for starting all wars is extremism (this might be the case in reality but there is no evidence in the passage to suggest this).

Answers B) and E) are not assumptions because neither is required to accept the passage's conclusion. The passage simply states that schools should educate children about the dangers of extremist politics to reduce the chances of future wars. Thus, none of them are *required* to be correct for the conclusion to be valid, and thus they are not assumptions. In addition, E) is a very bold statement, and there is nothing to suggest that the success of extremist parties will definitely continue to grow.

**Question 6: D**
In order to calculate when he should put in the potatoes, the first thing we need to know is how long the beef will take to cook.
- The question tells us that William is cooking for himself and 4 friends. Therefore this is a total of five people.
- He needs to allow 300g per person.
- He is cooking the meat medium rare in a **Hot** oven

Therefore, total weight of beef = 5 x 300 = 1500g

We know that in a hot oven, cooking time = 15 minutes per 500g plus 15 minutes and then an additional 10 minutes standing time.

Cooking time = (15 x 3) + 15 + 10 = 70 minutes

As we are cooking the potatoes in a hot oven, the table shows that they will be ready in 30 minutes. Hence, to ensure that both will be ready at the same, the potatoes should be put in 30 minutes before the beef is ready.

Therefore, we should put in the potatoes 40 minutes after putting the beef in, as they will both be ready after 70 minutes.

Hence, the answer is D).

**SECTION ONE**  2015

**Question 7: E**
The simplest way to find the answer to this question is by the process of elimination.
- If we look at the first condition it says the prize cannot be given to anyone who has been late more than twice. We can use this to eliminate the Grace first.

- The second condition is that the prize cannot be given to anyone who has failed to complete more than two pieces of homework by the set deadline.
  - Andrew had been set 60 pieces of homework and completed 56. Therefore he had missed 4.
  - Edward had been set 59 pieces and only completed 56, and so had missed 3, thus ruling him out of the prize.

- After the first two conditions have been satisfied, this leaves Carole and Ian left. Out of them, the one to receive the prize is the one with fewest non-A-grade pieces of work.
  - For Carole, 53/56 pieces of homework were awarded grade A, meaning only 3 were non-A-grade.
  - For Ian, 52/53 homeworks were awarded grade A, meaning only 1 was less than a grade A.

Therefore by following through the conditions, we can see that the prize will be awarded to Ian.

**Question 8: D**
This question requires us to read the information very carefully. It specifically asks us for a representation of plankton production in southern Polar Regions.
Firstly, the fact that we are in a polar region suggests that the variations throughout the year will be quite pronounced and easily distinguishable on the graph. Secondly, the fact that it is in a southern region suggests that there will be a surge in numbers of plant plankton in the months of December, January and February, as it would be summer during these months in the southern hemisphere. Using this information we can then see that the only graph to satisfy both criteria is D), hence it is the right answer.

Option A) this is the same graph for plankton production in northern tropical region so it cannot be correct.
Options B, C and E) The surge in plant plankton occurs during the months of July and August. In the southern hemisphere this would be during winter and so all 3 options must be incorrect.

## Question 9: C

The passage concludes that armed people are more likely to be shot than unarmed people. it cites the study conducted at the University of Pennsylvania as a main reason, which pretty conclusively showed that those who were armed had a greater chance of being shot.

Option A) is fairly irrelevant – whether the weapon is visible or concealed does not detract from the overall conclusion of the passage. Furthermore, option D) actually *strengthens* the argument if we accept this to be true, it gives a possible reason as to why those who are armed are more likely to be shot, because they are more likely to be involved in violent situations.

Options B) and E) neither strengthen nor weaken the argument, as they provide simple facts which provide additional information, yet do not counter the overall conclusion.

If we imagine that being armed is variable A and being shot is variable B, the principle of the argument states that A causes B. Hence C) does weaken the passage as it suggests that in fact, B causes A which reverses the whole argument. Thus, C) is the answer.

## Question 10: B

The passage says that recurring dreams contain messages that could potentially be very important; hence these dreams are repeated to ensure that you retain the information. However, there is no evidence to suggest that the explanations or information you receive in the dreams are true or beneficial. In fact, recurring dreams could simply contain useless information that is being repeated for an unknown reason. Therefore B) undermines the reasoning which supports the main conclusion of this passage, and so is the main flaw in the argument.

Answer A) is incorrect as we cannot completely discount supernatural explanations as having no value – the fact is that we do not know. Hence A) is not the main flaw. Answers C), D) and E) are irrelevant as they do not undermine the main conclusion of the passage – although they suggest that not all dreams are useful or that people can forget dreams, they do not discuss anything regarding recurring dreams (which is the main topic of the passage). Hence these are not flaws in the argument.

## Question 11: D

The passage discusses that buying expensive items and being materialistic do not necessarily make us happy; merely these actions are used as signal to show others certain qualities. The main argument of the passage states that in order to achieve true happiness, we should try to emulate our ancestors and devote more time to simple, traditional activities, so D) expresses the main conclusion of the passage.

Answer A) is a bold statement, which can only be, inferred from the first half of the passage, and so does not encompass a view of the overall argument. Answers B) and C) are not really supported by any evidence from the passage, whilst E) seems to be an intermediate conclusion, which then goes on to support the main conclusion given in D).

Hence, D) is the main conclusion.

## Question 12: B

When looking at the table, it is important to note that the same numbers of people were surveyed in each age group. Therefore this makes comparing between the different age groups much easier as we can say that that there were 100 people in each group. We can then solve the problem by considering each option in turn.

- Option A) we can see that Drink A is more popular than Drink B for all age groups except age 25-35. Therefore A) must be incorrect.
- Option B) the total number of people with a preference for Drink B = 20 + 31 + 33 + 19 = 103.
- Total number with no preference = 33 + 12 + 22 + 34 = 101. Therefore B) is the correct answer.
- Option C) we can see that 48% of people expressed a preference. Therefore 52% either expressed no preference or don't know and so C) must be incorrect.
- Option D) we can work out that the total number of people who expressed a preference was 103 (total for B) + 122 (total for A) = 225. As this is greater than 50% of the total number D) must be incorrect.
- Option E) we can prove this is incorrect by estimating an average of the percentage of people who did not know what they preferred. The Average = (19 + 19 +12 +23) / 4 = 18.25%, hence E) is wrong.

Therefore the only correct option is B).

## Question 13: D

This question seems quite difficult as we have two unknown variables to deal with, the standing charge and the amount per unit used. The best way to tackle this problem is to therefore solve it using algebra.

Let x = standing charge value and Let y = amount charged per unit

To work out x and y, we need to construct two equations and then solve them simultaneously. As the tariff could have been changed in any of the given months, we can only use May, June or July in our equations as the tariff must have been constant during these months. Using May and June as an example:

39.50 = x + 95y and 37.50 = x +75y

Solving these equations by elimination, 20y = 2

Therefore y = £0.10 and x = £30.00

Therefore we know that the standing charge is £30 pounds and the amount per unit used is 10p.

We can then check the remaining months to see when the tariff must have changed.

We can see that the tariff is consistent for all months including August.

E.g. £70 = £30.00 + 70 x 0.1  Therefore August is in the same tariff.

If we then check September we can see that it is the first month, which does not obey the same tariff. Under the old tariff, the monthly charge would be £30.00 + 80 x 0.1 = £38.00.

Therefore the tariff was changed in September and so D) is the correct answer.

## Question 14: E

When tackling this question, it is necessary to understand what both tables show. The first table suggests that children of parents working in professional or managerial roles tend to do better at GCSEs than those of manual labourers. The second table shows a different link between GCSE performance and type of school attended, with those in independent schools doing better than local authority schools. However it is important to note that although parental occupation and type of school affect success at GCSE level, there is no evidence that they are linked and so any connection between the two variable cannot be inferred from the tables, hence A) and B) are incorrect.

There is no evidence stated that one variable is more important than the other and so C) cannot be correct. Finally, although D) might seem plausible, it is not supported by any information provided and so must be incorrect.

Hence the only option which can be inferred is E) as both type of school and parental occupation clearly affect educational success in their own way.

## Question 15: A

This passage discusses the importance of learning at foreign language at school, as it increases one's confidence in communication and travelling. However the majority of British children do not study a foreign language, which could potentially limit their opportunities later on in life.

Option A) is a valid conclusion. There is a great emphasis on how British school children are not being taught a foreign language, and so the lack of education is holding them back in the business world. If we accept all the reason to be true, A) represent the correct conclusion of the passage..

C) is not a valid conclusion as the passage does not imply that the closing down of university language department has put teachers out of work, and therefore we cannot jump to this conclusion. D) Directly contradicts the passage – if we accept the reasoning in the passage to be true, we can see that there is a great need for Britons to learn a foreign language to succeed in the international business world.

B), and E) are both over concluding. The passage says that knowing a foreign language makes us more comfortable in interacting with foreigners. It makes no explicit claim about enjoying your holidays or losing out on business deals. Neither of these focus on the main point of the argument and instead make an indirect conclusion, hence they are incorrect.

## Question 16: B

The passage discusses how banning fun science activities have apparently led to a decreasing percentage of students enrolling on science courses at universities. The passage stresses that in order to increase the popularity of university courses; the regulations must be relaxed to allow exciting activities, as they rarely caused harm. This is directly based on the assumption that the 'exciting practicals' have a profound role in encouraging students to take up science. Thus, B) is an assumption from the passage.

Option C) actually disagrees with the passage, which states that some science practicals, even those that are minimally dangerous should be banned on health and safety grounds, so this is not an assumption. Option A) is also not an assumption of the argument, because the passage does not imply that the death of a few does not matter at all, rather than in the context of other factors such as the decreasing number of students taking science at university, this bears less significance. Hence A) is not the correct assumption as the main conclusion is not dependent on this being true.

Option D) is not an *assumption* because it actually strengthens the passage, as it provides further reasoning to relax health and safety regulations. Finally the passage does not necessarily imply that science can only be fun when dangerous and in fact many of the activities suggested are not dangerous at all, rather just more creative or interesting. Hence E) is not an assumption.

## Question 17: A

The passage emphasises how owning a home today is unaffordable for many people, as house prices have continued to rise due to increased levels of borrowing. This housing bubble created was a major factor, which contributed to the financial crisis, as it explained how homeowners were unable to repay their debt to pay back their loans. Thus, the main conclusion is that the government should prevent house prices from rising more to avoid a similar crisis in the future.

Answer B) actually weakens the argument, as it suggests that the passage is no longer relevant as house prices are no longer increasing, and so is incorrect. Answer E) also weakens this argument, as it suggests that house prices rising more than inflation rates is good for homeowners as it aides in paying back their mortgage, therefore undermining much of the reasoning in the passage. Finally D) also counters the argument, as it suggests that there is little that the government can actually do to lower house prices, therefore rendering the argument invalid.

Answer C) is fairly irrelevant, because it does nothing to help determine whether the government can do anything to prevent a continued rise in house prices, and so is incorrect. Therefore the correct answer is A) as it provides further reasoning why the government must act to reduce house prices – as the gap between prices and salaries has continued to widen, which suggests that home ownership is becoming even more unaffordable. Hence A) is the correct answer.

## Question 18: C

The best way to do this question is to work out the method yourself and then try and match it to one of the given answers. We know that out of 420 candidates, 210 will be sitting in the sports hall.
- Therefore we must share the remaining 210 students between the rest of the rooms.
- There are a total of 11 rooms.

The wording of the question here is very important: instead of asking for the same number of candidates in each room (which would be a far simpler calculation) we want to leave the same number of empty desks in each room.
- Hence, we must determine the total number of empty desks that will be left after all students have been accommodated for.
- Therefore, we need to work out the total capacity of the rooms
  Capacity = 26+20+24+28+24+20+24+16+20+18+26 = 246
- The total number of empty desks = 246 – 210 = 36
- Therefore, number of empty desks per room = 36/11 = 3.2727…

We can see that this method most closely agreed with C) and so this is the correct answer.

# SECTION ONE 2015

## Question 19: D

We can see from the question, there is a lot of key information that we need to take careful note of.
- There are a total of 80 people at the wedding reception.
- The marquee must allow a minimum of 3 metres squared per person.
- The couple want the most expensive chairs, and so they will be buying the gilt chairs.
- The couple want eight people per table, and so will be buying the round tables.
- The couple will purchase the complete package, and so the individual costs are irrelevant.

Firstly, we need to calculate the minimum area covered by the marquee.
Minimum area = 80 x 3 = 240 metres squared
Therefore, the smallest size marquee that the couple can order is the 9 x 27, as this will give an area of 243 metres squared. Hence, the price of the complete package is £1,385.
Next we must work out the price of the tables. As the couple need 10 round tables, cost = 10 x 6 = £60
Finally, the couple also requires 80 gilt chairs (as these are the most expensive). Cost = 80 x 2.40 = £192
Total cost = 1385 + 60 + 192 = £1,637. Therefore the correct answer is D).

## Question 20: E

The key to answering this question lies in looking at the results table and the graphs provided, and using the table to generate a scale for the x-axis of the graphs. After that it is possible to match each score to the correct bar of the bar chart.
The easiest way to start this question is to use the highest and lowest average marks on the test. The Lowest score was 20 by set 7MO and the highest score was 82 by set 7EV.
Looking at graph A, we know that the top bar must equal 20 marks. Therefore, this tells us that the x-axis is scaled in units of 20 marks each (each line goes up in 20 marks).
Therefore B) and C) must be incorrect as the top bar is greater than 20 marks. We can then use the highest mark achieved to discount graph A) as the bottom bar shows only 80 marks, whereas the highest average mark achieved was 82.
We then must compare graphs D) and E). The only clear visible difference between them is the fourth bar from the top, which should correspond to the mark 57. As in graph D) the bar for this mark is greater than 60, D) must also be incorrect. Hence E) is the correct answer.

# SECTION ONE  2015

**Question 21: C**

The passage states that certain illegal drugs interfere with a person's driving ability. The reasoning in the argument leads the author to suggest that driving with any level of any illegal drug should be a criminal offence, as this should ensure that driving ability is not impaired due to drugs. However, C) states that some legal drugs can also impair driving ability, and so as there is nothing in the passage to counteract this fact, this significantly weakens the argument. Thus, C) is the correct answer.

Answer A) is irrelevant to the main conclusion of the argument. Although some countries have legalised the drugs that impair driving ability, the problem still exists in this country. Answer B) is also not significant as the drugs referred to in the passage are illegal drugs – hence the possession or consumption of these drugs even if they do not impair driving ability is still a criminal offense.

Answer D) actually strengthens the passage as it emphasises how many drivers are unaware about the dangers of taking drugs and driving. Therefore, new laws to ban drug-driving as suggested in the passage should help eradicate driving under the influence of drugs. Finally E) does not represent the main flaw in this argument – yes, although some illegal drugs may improve driving ability, most others drugs (indeed the most common ones such as cannabis) have been proven to impair driving ability as is mentioned in the passage, and so E) is incorrect.
Hence C) is the answer.

**Question 22: A**

The passage reasons adding fluorine to water does not improve dental health, drawing upon the example that people who drink fluorinated water regularly do not have healthier teeth than those who do not. Therefore the reasoning in the argument is along the lines of: "Let us assume that A causes B. If you have A, then B will happen". A is drinking fluorinated water and B is having healthier teeth. However since this is not the case A cannot cause B.

Answer A) uses a similar reasoning to the one in the argument. Let us imagine that badgers living there is variable X and the presence of badger droppings is variable Y. It uses the assumption that X causes Y. Therefore if X happened, Y would happen. But since Y has not happened, X cannot be true. This most closely follows the reasoning in the argument and so is correct.

Answer B) follows the opposite pattern of reasoning. It claims that the presence of little piles of soil indicate that moles are present – the reasoning in the passage instead uses the lack of a particular feature to prove that something is not present. Hence this does not parallel the argument's reasoning.

C) warns that since foxes may kill chickens and are present in the area, the chicken house should be guarded against them. Therefore this introduces the idea of actually doing something, which is not mentioned in the argument.

Answer D) introduces the notion of ambiguity and doubt into the argument. Whereas the original argument uses the reasons to come to a definitive statement based on a logical fact. Answer D) suggests that the animal is probably a hedgehog simply as the person cannot think of anything else.

E) uses a slightly different reasoning to the argument, as it is along the lines of, "If A happened, B might happen. As A has happened, B probably has happened" where A is the presence of grey squirrels and B the absence of red squirrels.

Hence, the answer is A)

## Question 23: A

The issue raised in the passage is that people who live in towns with peculiar names find that their property values are lower. Therefore, they should feel able to rename their towns if it is in their best interests. This principle is related to whether the people of a community/town are being given the right to exercise their own will regarding their own towns. Answer A) illustrates this principle, so this is the answer.

Answers B) and E) both give the decision making ability to an external source (outside agencies and school staff) which is in contradiction to the principle of the passage.

Answer C) is completely irrelevant, discussing on what basis officials should be elected, but not by whom. Answer D) disagrees with the passage, as nowhere in the passage does it state that new residents cannot expect to be part of the local community. Therefore, D) is also incorrect.

## Question 24: B

The fastest way to calculate the answer for this question is to see how much personal allowance and married couple's allowance Mary is eligible for and then calculate what her taxable income is.
- Looking at the column 2000-01, as Mary is aged under 65, her personal allowance is £4385.
- The table shows that the elder spouse (Mary' husband) is aged Under 65 too, she is not eligible for any married Couple's allowance.
- Her total income is £5,585

Therefore taxable income = 5585 – 4385 = £1200

As this amount is within the 10% tax band, she must pay 10% of £1200. Hence the total tax paid = 120/10 = £120

Hence, B) is the correct answer.

## SECTION ONE  2015

**Question 25: D**
The fastest way to answer this question is to work out the total rainfall between the beginning of June to the end of September and then divide by the number of months (which in this case is 4).
We can see from the graph that at the end of September the cumulative rainfall was 400mm. At the beginning of June, cumulative rainfall was approximately 180mm.

Therefore total rainfall in these months = 400 – 180 = 220mm
Average monthly rainfall = $\frac{220}{4}$ = 55 mm. The closest answer is 57mm/month and so D) must be the correct answer.

Note, we know that although C) is quite close to our answer, it cannot be correct as we can see the total rainfall in these months it clearly over 200mm and so our average must be greater than 50mm/month.

**Question 26: B**
We are told that the bath has a capacity of 360 litres.
The hot tap can fill it in 15 minutes.
Therefore, the rate of filling for the hot tap = 360/15 = 24 litres/min
The cold tap can fill it in 10 minutes.
Hence the rate of filling for the cold tap = 360/10 = 36 litres/min

In the first 1.5 minutes, both taps are running. Therefore the amount of water in the bath = $(1.5 \times 36) + (1.5 \times 24) = 54 + 36 = 90\ litres$
We can use this information to work out the scales of the axes on the graphs. The x-axis must go up in units of 1.5 minutes each and the y-axis must go up in units of 90 litres each.

As we are told that the bath is left ¾ full, this would mean that the final volume is 270 litres. Hence, C) and E) must be incorrect as the total volume of water in these graphs is 360 litres.
After 1.5 minutes, we are told that the cold tap is switched off leaving only the hot tap still on. The amount of water left to fill up = 270 – 90 = 180 litres.
Time taken to fill remainder = 180 / 24 = 7.5 minutes.

Therefore, the bath would be at the correct level of 270 litres after a total of 9 minutes. As we know the scale of the graphs, this would correspond to the 6$^{th}$ line along the x-axis. Hence B) is the correct graph.

## Question 27: B

The passage describes the potential dangers of storing personal data on computers on our privacy, and explains how the Data Protection Act ensures that this information is kept confidential except in extenuating circumstances. It also mentions that, as there is a possibility of this information being misused in the future, we *must not allow the principle to be abandoned* (referring to the Data Protect Act). All the reasoning given ultimately supports the idea that the Data Protection Act should not be abandoned for our own safety. Hence B) is the main conclusion of the passage.

Answers A), C) and E) are all reasons in the passage which support this conclusion, and are not themselves conclusions. Answer D) is irrelevant to the overall conclusion and we cannot be sure to what extent the Data Protection Act might potentially interfere with crime detection, hence D) is also incorrect.

## Question 28: C

The passage describes how politicians use sound bites to justify their proposed actions as it allows them to take advantage of the sentiment of the public and so get away with their decisions.

Answer A) is not a valid conclusion. The second line clearly states that sound bites are used to justify a politician's action, rather than use a sound bite to avoid a proper explanation.

Answer B) is not a valid conclusion because the passage does not state that politicians always believe that their actions are right, rather than the sound bite is a tool to justify their actions, even if they are unsure whether it is right or wrong. Thus, we cannot *reliably* conclude that B) is true.

Answer D) is not a valid conclusion because the passage directly contradicts this statement. The passage suggests that sound bites are used to appease the public or quieten public discontent; rather it is only the author of the passage that seems alienated. Hence D) is incorrect.

Answer E) is not a valid conclusion because no reference is made in the passage to whether politicians actually prefer to use sound bites rather than provide a proper rationale. Instead the sound bite could be a last resort that politicians use if they cannot properly justify their actions. Notice how A) is extremely similar to E), and so in this case is seems likely that both answers are wrong.

Answer C) can be reliably concluded. We are told that the use of sound bites in many ways is immoral and is effectively a way of manipulating the public. We also know that instead of sound bites, the moral thing to do for politicians would be to accurately explain the reasoning behind their actions. Hence C) encompasses the whole argument and so is the main conclusion.

## Question 29: D

The passage discusses how the British Constitution has greatly benefited over the years by not being too rigid and following logic, but instead being guided by following common sense. It concludes that the reason we have avoided several problems in the past is because we have always followed common sense and never been guided by logic. However, it is not necessarily the case that using logic in the past would not have avoided problems wither, as we have never tired using it and so do not know what would have been the consequences. If this is not the case, the passage's conclusion that ignoring logic and following common sense is far better do not necessarily follow from its reasons, so Answer D) correctly identifies an assumption in the passage.

Answer C) strengthens the passage's conclusion, if true. However, it is required for the passages' conclusion to be valid, so it is not an assumption, as it is not an integral part of the argument.

Answer A) is not an assumption, because it directly contradicts the passage. The opening line states how logicians did not devise the British Constitution, and so A) is incorrect. Furthermore B) is also not an assumption, as there is nowhere in the passage which implies that the British Constitution even today is written down, and so it would be wrong to assume this in the argument. Hence B) is not correct.
Answer E) is not an assumption because although the passage discusses the benefits of ignoring logic and following common sense, this is a great over conclusion of the argument. There are several advantages of using common sense, but we cannot assume that it must always be right to ignore logic to produce the best result. Hence answer E) cannot be an assumption of the passage.

## Question 30: D

The best way to do this question is to work out which teams cannot mathematically win the league.
Firstly we need to calculate the total number of games played by each team. There are 6 teams in the league and we are told that all teams played each other 4 times. Therefore total games played = (6-1) x 4 = 20 games. We can then start from the bottom of the table to work out which teams cannot win.

Starting with Bottom Albion. We know that they only have 5 games remaining. Therefore, the maximum number of points that they can win with 5 wins = 5 x 3= 15 points. This would give them a total of 30 points, meaning that they would still be below East Rovers, and so they cannot win the league.

Although Top Town, are second from bottom, they have 7 games remaining. Therefore, the maximum number of points they can achieve is 17 +21 = 38 points. Provided East Rovers lose or draw the rest of their games, Top Town could mathematically still win the league.

South United have 3 games remaining and so could only score a maximum of 31 points which would make them level with East Rovers. We are told that in this case, the team with the most wins will win the league. Even if South United win all of their games, they would only have a total of 7 wins, which is less that East Rovers, and so they cannot win the league.

Although North Rangers also have 22 points, with 3 wins out of 3, they could score 31 points in total with 9 games won (more than East Rovers if they lose all their remaining games). Hence North Rangers can win the league.
It is clear that West Athletic and East Rovers can win the league, which means that only Bottom Albion and South United cannot win. Hence only 4 teams can win the league and so D) is the correct answer.

**Question 31: C**
This seems like quite a long question as there are 9 different classes to choose from, and so to calculate the price per hour for each class would be quite time consuming. Therefore, the quickest way to do this question is to eliminate answers by working out which do not satisfy the conditions stated before working out the price per hour.

The first condition states that she does not want to leave home before 09:15 and return by 16:15.
- We can therefore eliminate Painting Class 3
- We can eliminate Dancing Class1 and Class 2
- We can eliminate Pottery Class 1

We must then work out the price per hour of the remaining classes
- Painting Class 1 = 30/2 = $15 per hour
- Painting Class 2 = 40/1/5 = $26.66 per hour
- Pottery Class 2 = 40/2 = $20 per hour
- Writing Class 1 = 35/2.5 = $14 per hour
- Writing class 2 = 24/1.5 = $16 per hour

Therefore she will choose writing Class 1. The price of this class is $35 and so C) is the correct answer.

## Question 32: D
This question is difficult to describe on paper, as it requires visualisation of 3-D shapes. When looking at the remote from the direction of the arrow, you would see a row of five small buttons, regardless of any of the larger buttons.
Therefore the answer is D as the small button in the first column is missing and so cannot be a possible view.

## Question 33: C
The passage discusses how the apparent fall in crime rate in many towns and cities such as Kings Lynn is thought to be due to the implementation of CCTV cameras, and so they have been a great success in controlling "opportunistic" crime. However C) greatly weakens the association between CCTV implementation and reduced crime as it introduces a confounding variable: extra police presence, which may have caused the fall in crime rather than the CCTV cameras. Thus C) is the correct answer.

Answers A) and E) are not relevant as the passage simply states that CCTV reduces crime rates. Whether they are recording or not, or can be used in court convictions is not important to the overall conclusion.

Answer D) seems irrelevant as it does not counter or weaken the effectiveness of the CCTV cameras. Even if they are an overreaction, they have produced great results and so this does not oppose the main conclusion. Finally B) could actually be seen to strengthen the argument – as even though the cameras are situated in the centres of towns and cities, they still great reduce crime rates (probably because so called "opportunistic" crime is greatest in the busy city centre).
Hence C) must be the correct answer.

## Question 34: A
The passage describes how learning a foreign language at school is essential if people want to work at the highest level. The argument stated here is along the lines of, for X to happen, Y is required – where X is working at the top level and Y is learning a foreign language. The argument then suggests that is people have Y, then X will happen. Therefore by replacing the variables with X and Y, we can see how the argument has treated a necessary condition for success as a sufficient condition. Just because A is necessary does not mean that if you have A, B will definitely happen. Hence A) is the correct flaw in the argument.

Answer B) does not identify a flaw in this passage, because it is the opposite argument to what the passage is actually suggesting. For business success, learning a foreign language is necessary, not sufficient – therefore it disagrees with the passage.

Answer C) does not directly relate to the discussion as the passage does not in fact include any main examples, and rather just states assertions that we should accept to be true. Therefore it is not the main flaw in the argument.

Answer D) can be seen as a flaw in the passage as the author assumes that learning a foreign language will directly help students succeed in the business world. However due to the reasoning given in the passage, which we assume to be true, there is no evidence to suggest that this relationship is not causal. Furthermore, this small flaw is outweighed by the much larger flaw in A) and so in not the *main* flaw of the argument.

Answer E) also somewhat is irrelevant. Nowhere in the passage does it mention the past; rather the passage suggests that this is a current problem which should be dealt with now to prepare for the future.
Hence the answer is A).

**Question 35: B**
The passage criticises banks, which pay large bonuses to their employees, saying that it cannot be justified. This statement at the beginning of the passage is then supported by several reasons, which argue that it discourages young people from taking jobs in the industrial sector and also promotes adverse risk taking with shareholder's assets. The remainder of the passage explains why firms pay large bonuses, but then again highlights how the cost for these can fall on to the consumer. Therefore, the main conclusion of the passage must be B).

Answers A) and C) identify reasons given in the passage, which go on to support the conclusion given in B).

Answers D) and E) could be seen as conclusions from the passage, but both would require leaps of logic not made in the passage in order to be conclusions, and thus are somewhat criticisable as conclusions.

Also, if we accept either of these answers as true, they both go on to support the conclusion given in B), making them *intermediate* conclusions, not the main conclusion.

## SECTION ONE     2015

**Question 36: C**

First we calculate the total time available to sign autographs in minutes:
- There are 9 hours available between 9am and 6pm.
- This means that there are 9 x 60 = 540 minutes to sign autographs.

Next we are told that he can sign ten times per minute and has to have a ten-minute break after each hour. Therefore the easiest way to do this is to find out how much the organisers can earn each round of 70 minutes.
- Amount earned per minute = £10
- Amount earned per hour + 10 minute break = 10 x 60 = £600

Therefore we can split the 540 minutes available into 7 rounds of 70 minutes and a remainder of 50 minutes.
Therefore total amount raised = (7 x 600) + (50 x 10) = 4200 + 500 = £4700
Therefore the answer is C).

**Question 37: D**

First we need to calculate how many cubic metres of mulch are required for each plot of land.

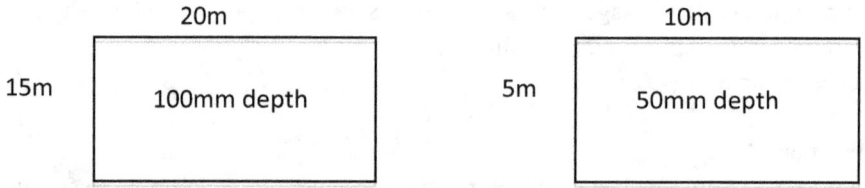

Area of Plot 1 = 10 x 5 = 50 metres squared
If we then look at the table, underneath the 50mm depth column, we can see that we would require 2.5 cubic metres of mulch.
Area of Plot 2 = 20 x 15 = 300 metres squared
If we then look at the table underneath the 100mm column, we can see that 100 metres squared requires 10 cubic metres of mulch.

We can then extrapolate this, so that 300 metres squared requires 30 cubic metres of mulch.
Total mulch required = 2.5 + 30 = 32.5 cubic metres
Total weight of mulch = 32.5 x 0.65 = 21.125 tonnes
(NB: This is a hard sum to calculate without a calculator. A simpler way to do this is to work out 50% of 32.5 = 16.25. You can then increase this by 10% and then 5% to work out 65%)
Therefore, we would require 22 one-tonne bags to ensure we have enough to complete the mulching of the two plots. Hence D) is the correct answer.

**Question 38: E**
This is a question that involves many steps to find the correct answer. Firstly, we are told that the time is supposedly nine minutes past ten.
➢ Therefore, the first thing we must do is work out how much time has passed since he reset his watch and looked out again across the road this morning.
➢ He reset his watch at 22:05 and again looked out at 10:09. Therefore, we can see that the time passed was 12 hours and 4 minutes.
➢ Next we can work out the actual time when he reset his watch last night. As the reflection showed a time of 22:05, this means the actual time was 20:55.
➢ Therefore, the actual time this morning must be 12 hours and 4 minutes after 20:55. Therefore the actual time is 08:59.
➢ Hence, when he looked across the road, he would see a reflection of the time 08:59.

Therefore, E) is the correct answer, as it is the correct reflection of the time 08:59.

**Question 39: C**
This passage highlights how many consider that homeopathic medicines work solely due to the placebo effect, and so are only effective as patients believe they are useful. This stance is countered by a particular study in cows which tested the effects of homeopathic medicines compared to a placebo, and found that there was a difference in beneficial effects between the two. Therefore, if we accept the study to be fairly conducted and the result true, the most logical conclusion from this is that the beneficial effects of homeopathy in cows cannot be due to the placebo effect, so C) is the right answer.
Answer A) contradicts the findings of the experiment as the study showed more cows developed mastitis with a placebo than the homeopathic remedy, and so the effects of homeopathy cannot be solely due to the placebo effect, hence A) is incorrect. B) Criticises the validity of the experiment conducted, however there is no evidence to suggest that we should discount the findings of the study with the cows, and so we cannot conclude B).
Although in this case, the experiment was carried out on cows, D) seems to be an over conclusion as we cannot be sure that homeopathy is only beneficial to animals. If humans or other species were tested, they may see the benefits of homeopathy. Finally E) could be described as an intermediate conclusion, as it goes on to support the notion in C). Although the study does suggest that there are several beneficial effects of homeopathy, there is no further evidence we can use to extrapolate these results to humans and other species. Furthermore, the sample size maybe too small and so these results maybe unreliable. Therefore much of the criticism maybe justified and so we cannot conclude E).
Hence C) is the correct answer.

**Question 40: C**
This passage emphasises that insurance must be compulsory for dog owners, arguing that it is unfair that victims must take the owner to court to receive compensation, which can be an arduous and expensive process. This would mean insurance companies pay for injuries caused as they do for damage to cars and homes.

Looking at the reasoning used, we can see that the argument stresses that just because a law being passed would make dog owners have to take out insurance for their dogs, it does **not** mean that they would do it. If this is the case, then the passage's reasoning no longer supports the conclusion that all injuries would be paid for by the insurance companies. Thus C) is a valid assumption of the passage.
Answers A) and E) both weaken the argument. A) Suggests that if the law was passed, dog owners may not be able to afford the insurance and so may not opt to buy it and their dogs would not be covered by insurance. Similarly E) weakens the reasoning to make dog insurance compulsory as it suggests that injured victims can easily take owners to court and receive compensation. Therefore both of these are not assumptions.

Answer B), if true, would reinforce the conclusion, providing additional strength to its reasoning. However, it is not *required* to be true for the conclusion to be valid, so this is not an *assumption* of the passage.

Answer D) is irrelevant to the overall conclusion of the passage. The fact that dogs are illegal or legal makes no difference to whether passing a law to make insurance compulsory would ensue that injuries are funded by the insurance company, rather than the taxpayer.

Hence, the answer is C).

## SECTION ONE 2015

**Question 41: D**

The passage discusses how there is a disproportionate number of students who study at elite universities to those who attend private schools. This increased proportion is seen as linked with better teaching in private schools, which allow students to achieve better examination results and therefore gain admission into better universities. Thus, it concludes that the teaching in private schools must be better than state schools. Note that we are asked for the answer that would **most weaken** the argument – so we may see multiple answers which would weaken the argument and must choose the one with the most direct link as the strongest.

Answer E) would *strengthen* the passage. It suggests that as private schools accept students from poorer backgrounds and still are better at getting their students into better universities, family background does not really determine or affect the grades students achieve, rather the better performance of those in private schools must be down to the better education they receive.

Answers A) and C) are irrelevant. The passage does not refer to whether students in private schools do better in just A-levels, but that they achieve higher grades overall. Answer C) also suggests that a majority of those in Oxbridge are from state schools. However this number of students in private schools (7%) is still disproportionately small to those in Oxbridge (40%), so this in fact strengthens the argument. It is adding no new information that cannot be deduced from the question.

Answer B) does weaken the passage, as it suggests that family background and wealth might be the reason more private school students go to elite universities, rather than better education. The implication is that factors other than the school's performance may influence attainment. However there are a few missing steps between this background and A-level results, so it is not a very strong argument.

Answer D) is the strongest and therefore the correct answer, as if fewer state school students sit A-levels, this could be the reason why fewer get the opportunity to go to Oxford and Cambridge. In other words state school exam results might not be worse, but the good results might be enjoyed by fewer pupils. Now this is not a wholly satisfactory answer as the underlying reason behind fewer people sitting A-levels could be a lower standard of teaching and attainment in previous years – but as it is a solid and direct reason why despite comparable teaching fewer state school pupils might make Oxbridge selection it is the strongest answer to the question.

**Question 42: C**
We know that the examination contains five questions:
1                2                3                4                5
To make the question simpler, we can split the 5 questions into two groups, which allow us to test each condition individually.

Group 1:         1                2                3
For this group, we must choose at least one of the first two questions. Therefore the combinations allowed are:
1                2                3
1                2
1                3
2                3

Group 2:         4                5
For this group, we must choose not more than one. Therefore, we can either choose question 4, 5 or indeed none of them.
Now that we have established which questions we can pick from each group, and knowing that we must pick choose 3 questions in total, the possible combinations to choose are:
1                2                3
1                2                4
1                2                5
1                3                4
1                3                5
2                3                4
2                3                5

Therefore the total number of combinations possible is 7 and so C) is the correct answer. If we had not split the initial five questions into 2 groups, then it is more likely that we would have missed the combination of questions 1,2,3 and so would have incorrectly chosen answer B).

# SECTION ONE — 2015

**Question 43: E**

To answer this question, we must calculate the total time taken for each of the possible journeys – although some options may take less time, it is important to check that they satisfy Hester's conditions and she has enough time to spend in each town before the next train.

**Option A:**
- Leave Birmingham at 09:00 → Arrive at Tamworth at 09:30
- Spend 1 hour in Tamworth until 10:30
- Catch 11:30 train → Arrive at Derby at 11:50
- As she must spend 1.5 hours here she cannot catch 12:15 train

**Option B:**
- Leave Birmingham at 10:00 → Arrive at Tamworth at 10:28
- Spend 1 hour in Tamworth until 11:28
- Catch 11:30 train → Arrive at Derby at 11:50
- Spend 1.5 hours until 13:20
- Leave Derby at 14:00 → Arrive back at 15:00
- Total time is 5 hours

**Option C:**
- Leave Birmingham at 11:00 → Arrive at Tamworth at 11:30
- Spend 1 hour in Tamworth until 12:30
- Catch 12:45 train → Arrive at Derby at 13:05
- Spend 1.5 hours until 14:35
- Leave Derby at 15:20 → Arrive back at 16:12
- Total time is 5 hours and 12 minutes

**Option D:**
- Leave Birmingham at 12:00 → Arrive at Tamworth at 12:45
- Spend 1 hour in Tamworth until 13:48
- Catch 13:45 train → Arrive at Derby at 14:00
- Spend 1.5 hours until 15:30
- Leave Derby at 16:10 → Arrive back at 16:55
- Total time is 4 hours and 55 minutes

**Option E:**
- Leave Birmingham at 13:30 → Arrive at Tamworth at 13:45
- Spend 1 hour in Tamworth until 14:45
- Catch 14:45 train → Arrive at Derby at 15:05
- Spend 1.5 hours until 16:35
- Leave Derby at 17:00 → Arrive back at 18:15
- Total time is 4 hours and 45 minutes (shortest time taken)

## Question 44: D
Again this question is difficult to describe on paper, as it requires visualisation of 3-D shapes. The best way to prepare for this question is to practice with dies, as these types of questions are common in this exam. Alternatively, you can use cut-out nets in the exam to make the die if you have enough time (remember you can always take scissors into your exam!).
Without a cut out net, this question is very difficult. The best way to tackle it is to first be sure of exactly which numbers are on the sides you cannot initially see.
➢ Opposite the 5 is a 2
➢ Opposite the 3 is a 4
➢ Opposite the 1 is a 6

Using this information, you must look at each of the options and attempt to do this question by elimination. But if you do have time, try to make a cut out net, as it will make the question much easier.

## Question 45: C
The passage argues against the claim that you can work out what a particular sentence in a book means by simply asking author or guessing what he intended to mean, as the true meaning of a sentence will inevitably depend on your own interpretation of it. If we accept all the reasons given in the passage to support this claim, we therefore find that C) expresses the main conclusion in this passage as it encompasses the whole argument.

E) is also a valid conclusion from the passage, but it goes on to support the statement in C). Thus, E) is an intermediate conclusion in this passage, *not* the main conclusion. A) Could also be seen as an over conclusion, but it seems too far stretched that no one can determine what the sentence means, as one can work out what the sentence means to them. Answers B) and D) are reasons given in the passage to support its conclusion.

## Question 46: C
The reasoning in this passage can be described, as "For A to happen, B is required. Since A did not happen, you did not have B." Here, "A" is getting a job, and "B" is writing a **good** application letter.

Answer A) could be described as "If you have B, then A will definitely happen. Since A did not happen, you did not have B", with A being having a place in the final and B recording the best time in the heat. This is not the same as in the passage as you did not need to write the best application letter to get the job - the use of the superlative means A) must be incorrect.

Answer B) reasons as "For A to happen, B is required. You don't have B, so A will not happen". This is not the same as in the passage, so B) is incorrect.

Answer C) reasons as "A cannot happen, with B. Since A did not happen, you did not have B". This most closely parallels the reasoning in the argument; hence C) is correct.

Answer D) reasons as "A will only happen if B has happened. As B has happened, A will happen", where "A" being promoted and "B" is working hard. This is not the same as the passage so D) is incorrect.

Answer E) reasons as "For A to happen, B is required." However it then says that as B has nearly happened, A should happen. This is not the same as in the passage.

**Question 47: A**
The passage discusses how the credit crisis as well as previous government policy decisions has caused many pensioners to be at risk of poverty. It claims that, as we have had a pension crisis looming for years, which the government is partly responsible for, it should be willing to face the consequences of its actions and act to restore them. Thus the principle underlying the passage is that since it was one's fault that an error occurred, they should be the ones to try and correct their mistakes.

Answer A) follows a similar principle, in which the person could have avoided dropping plates had he not carried so many (similar to the pension crisis which could have been avoided by the government taking better decisions). Furthermore, as he dropped the plates and is responsible, he feels compelled to replace the broken ones, thus making up for his actions.

None of the other answers refer to both notions that if better decisions were taken beforehand, the mishap could have been avoided **and** the fact that it is the responsibility of the one who is culpable to redeem their actions, and so these are not the principles in the passage.

# SECTION ONE 2015

## Question 48: A

To solve this problem, we must first calculate the how many containers are used each quarter (from Jan-March, Apr – Jun etc.)
- Quarter 1 = 2000 + 3000 + 1000 = 6000
- Quarter 2 = 3000 + 2000 + 2000 = 7000
- Quarter 3 = 1000 + 1000 + 3000 = 5000
- Quarter 4 = 1000 + 2000 + 3000 = 6000
- Therefore the total usage in the year = 24000 containers

We are told that the takeaway currently has 2,000 containers left. Therefore the total amount that needs to be delivered is 22,000. We can use this information to then eliminate any of the answers, which do not deliver 22,000 containers.

D) is incorrect as the total number of containers delivered is 24,000 which would leave a remainder of 2000

B) And C) must both be incorrect as we are told that the maximum number that can be delivered at one time is 6,000 and for both, it says 7,000 will be delivered in quarter 2. Therefore these two options are wrong.

We can then compare answer A) and E). Although both have a total of 22,000 containers delivered, we must consider each quarter individually.

For E) 6,000 containers need to be used in Quarter 1. Therefore, the takeaway would have to use all 4,000 delivered and the 2,000 remaining. However, this means that they would not have enough containers to meet the usage for quarter 2 (only 6,000 containers are delivered and the usage is 7,000). Hence E) must also be incorrect. Hence A) is the correct answer.

## Question 49: C

The quickest way to do this question is by the process of elimination, starting with the lowest amount provided in the question, A) 58p and then working upwards. For each amount provided, it is necessary to see whether it is possible to make up the amount with less than 3 stamps, and so, if we find an amount for which it is not possible to do so, that will be the correct answer.

For A) we can make up 58p = 56p stamp + 2p stamp. Therefore A) is incorrect.
For B) 65p = 56p + 9p. Therefore B) is also incorrect.

For C) we cannot use only two stamps to make up 68p. A possible combination could be 62p + 5p + 1p although there are several others available. Thus C) is the correct answer.

It is important to note that it is not possible to make D) 75p with two stamps either, and so if you miss C) then the only other possible answer would be D). This highlights how starting from the lowest amount first and then working upwards could avoid choosing this answer.

For E) 79p = 62p + 17p. Therefore E) is incorrect.

**Question 50: E**

The best way to tackle this question is to imagine taking photographs from different points on field and then writing down the order or trees you would see from left to right.

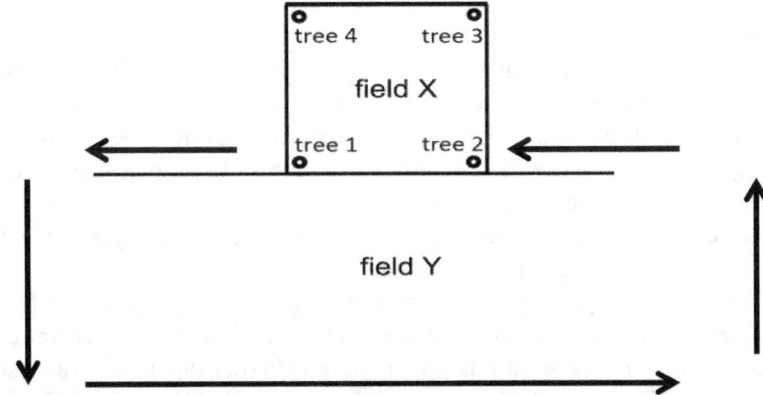

To ensure that you find all the possible combinations, you should start on the edge of Field Y at the marked point and then work around the perimeter of field Y noting every time you find a different combination. Therefore if we start at that point and work round in the direction stated the orders seen are:

➤ 4 – 3 – 1 – 2
➤ 4 – 1 – 3 – 2
➤ 1 – 4 – 3 – 2
➤ 1 – 4 – 2 – 3
➤ 1 – 2 – 4 – 3

Therefore as there are 5 different orders from left to right, the correct answer is E).

**END OF SECTION**

# Section 2

*Would society be better if there were more scientists in positions of political power?*

This is a difficult question in that it requires knowledge of both politics and science, and the interaction between them. It can be made simpler by clarifying in the opening paragraph the manner in which you are going to make the argument – i.e. from the points of view science can benefit society. Nevertheless, it remains a challenging question.

### Introduction
- Begin by defining your terms, as above. Indicate what you consider to be 'political power', and whether in this context it refers to positions in government. It is important to have an opinion regarding the question from the beginning – it is advisable to argue one side and acknowledge the other side of the argument later in the essay.
- Clearly state your opinion within these defined parameters, and outline the main arguments you'll be discussing.
- There are many ways in which society can be improved, which could be categorised into economically, socially, culturally etc. Therefore when addressing the question it is important to acknowledge how diverse society is and therefore state which aspects of society you will be focussing on.
- From an economic point of view, society might benefit, as scientists may potentially be able to allocate resources more efficiently, whilst also devising an efficient economic plan to aid the country. In contrast, from a social, and possibly a more realistic point of view however, the reality might be different.
- A key issue at the heart of this question's concerns is to evaluate the main qualities of a scientist – and then use this to either suggest why/why not they would be used to benefit society.

*Potential arguments why society would be better:*

- You could make an argument that a major aspect of science is using research and empirical evidence to make informed decisions. This would be of great benefit to society, for example in allocating resources to particular sectors or making decisions, as one could argue that a scientist would consider all the possible variables to arrive at the most logical conclusion. For reasons like this, it is often beneficial to give an example, e.g. how much money to allocate to the NHS.
  - **Counter-argument**: It is worth noting that not all arguments can simply be made on factual evidence. Furthermore, when confronted with a new problem, the time needed to research the problem may not be available. Hence a scientific approach may not always be the most practical approach. Give an example of a situation where this may be the case.

- Having more scientists in political power may lead to a more balanced government. Currently, the majority of politicians are from a political, economic (humanity based subject) background. If there were more scientists, this may help to even out the demographics of parliament so that it more closely represents the demographics of the country, rather than simply being composed of political graduates.

- If more scientists are in political positions, this may help to increase the popularity of scientific subjects at school and the numbers of those who choose to study science at university. It is worth mentioning here what science has achieved in the past: greater healthcare, discovery of antibiotics, extensive space exploration to name a few. Increased scientific interest in society therefore will inevitably lead to greater discoveries in the future.

- Finally, with certain aspect of society such as health, energy and the environment, a scientist is more likely to understand the consequences of new policy better than a non-scientist as it relates to a subject that the scientist has expertise in. Therefore they would be better placed to decide issues in government.

*Potential arguments why society would be worse:*

This is arguably the easier side to argue for this particular question. When stating why society would be made worse, there are several opportunities for counter-arguments which themselves can be further challenged (something that will make the essay much better).

➢ Although scientists may excel in their particular field, it needs to be mentioned that many will have had no experience governing or leading a country before. To be in a position of political power, we need to consider all the other qualities that are required: being a good listener, communicator, leader etc.

➢ One of the biggest dangers of promoting more scientists to political power is the fact that several atrocities in the past have been allowed under the banner of science. This provides an excellent opportunity to cite meaningful examples. For example, the rise of science in society might lead to eugenics – the selective breeding of people for the so-called benefit of society, or even the rise of Nazism and the pursuit of the pure Aryan race. Therefore it is necessary to consider the possible social implications of promoting more scientists to political power.
  o Counter-argument: there is always an argument to be made that no one is perfect and that arguably scientists have done more good than bad. The great number of scientific discoveries have saved more lives than were taken due to absurd scientific theories.
  o Counter-counter argument: we can mention how several scientific discoveries have the capacity now to completely ruin the society in which we live. An example of this is the discovery of nuclear fission and the atomic bomb, which could potentially destroy modern society.

➢ Another point to mention is that increasing the number of scientists to political power, especially in other more religious countries, might increase tensions between religious and secular communities in society. It is worth noting that several societies disagree with the theory of evolution and see science as a means of destroying their beliefs. This is a controversial topic and so should be discussed with caution, but is worth mentioning.

*Time lag*
Another angle of assessing the potential benefit to society is through timeframe – and evaluating what would be the short and longer-term consequences of having more scientists in political power. An argument therefore could be made that focuses more on advancement's long term impact rather than immediate impact.

*Conclusion*
- Make sure that you include potential counter arguments for any of the above points that you make – it is vital that your essay comes across as balanced and not one-sided.

- In your conclusion you could argue that there would be certain benefits to society, but at the same time other ways in which society might suffer.

- Another distinction you could make in your conclusion is between how we ought to evaluate the overall benefit to society and whether the economic or social argument is more important. Would it be right to allow great scientific advancement whilst at the same time allowing for the deterioration of the arts sector in society? Exploring these political implications is a good way to end the question.

*Is it possible to justify abortion without justifying infanticide of recently born babies?*

Before answering the question, it is important to fully understand what the question is actually asking. This is not a debate for /against abortion – instead whether abortion and infanticide are equally justifiable or fundamentally different. In what ways are they similar? How are they different? What are the moral and ethical objections of each? Before beginning to answer, consider these questions, as they will form the basis of your answer. For example, you might argue that infanticide cannot be justified despite allowing abortion due to the fact that a living baby is tangible being, but a foetus (or embryo) is still an intangible object.

*Introduction:*

- Begin by defining your terms, as above. Indicate what you consider to be 'abortion' and "infanticide" – it does not matter that you know the actual medical definitions, rather a sensible understanding of the topics at hand.
- Though it may seem obvious, it is worth briefly discussing *why* abortion and infanticide occur.
- Clearly state your opinion within these defined parameters, and outline the main arguments you'll be discussing.

*Potential arguments to justify abortion without justifying infanticide:*

- You could make an argument abortion and infanticide are fundamentally different. In the case of abortion, since the foetus has not been born it is not considered to be a person and therefore should have no rights. Hence the decision whether to keep or terminate the fetus' life should lie with its mother. In contrast, once the baby is born, infanticide could be deemed as murder as the baby is legally a person and so has fundamental human rights.
    - **Counter-argument**: it is possible to argue that all forms of life, whether born or still in the womb is sacred, and so both abortion and infanticide are morally wrong.
- Abortion could be justified under the grounds that an embryo is not a developed enough being to feel pain as its nervous systems or organs may not have formed yet. Therefore terminating the embryo's life at this stage would not cause any hurt to the baby. In contrast, once a baby has been born, its senses have developed to feel pressure, and pain and therefore infanticide would cause hurt to the baby.

- ➢ The risk to women during childbirth is substantial especially in less well-developed countries, and therefore abortion (particularly in the earlier stages) can be seen as a safer method of terminating the pregnancy if the child is unwanted – rather than waiting till after the child is born giving risk of birth complications. This would provide an argument to justify abortion without necessarily allowing infanticide.
- ➢ There is a strong argument to be made that abortion in this country at least, is justifiable as it is only allowed if the pregnancy was unwanted or if the baby has a genetic defect (rather than terminating after finding out the baby's sex). However, in other countries, the practice of female infanticide is common, as boys are preferred in these male-dominated societies. Hence, since the reasons for abortion are more justified than for infanticide, abortion can be allowed.

*Arguments why abortion cannot be justified without justifying infanticide:*
- ➢ All forms of life, whether born or still in the womb is sacred. This can be supported by considering duty-based ethics, and suggesting that we all have a moral obligation to preserve life in all forms. Hence there is no ethical difference between abortion and infanticide, and both cannot be justified.
- ➢ In both cases, it is possible to argue that it is not the child's own decision as in neither case, the child is competent enough or able to state whether he wants to live. On this note, we can argue that abortion and infanticide are both not acting in the best interests of the child and so both should not be justified.
  - o **Counter-argument**: we can argue that since a foetus isn't a person, it should not have the same rights or deemed to have any interest of its own. Therefore it should be legally and ethically moral that this decision can be taken by the mother.

- ➢ There is a strong argument that infanticide and abortion are effectively the same phenomenon, as they both represent ways of ending a baby's life – abortion can be thought of as the modern solution to terminating pregnancy through the use of technology which is often not accessible or unaffordable for many in poorer countries. In contrast, infanticide provides a more practical, accessible option to end a baby's life. If we assume that there are no moral objections to ending a baby's life, then it can be argued that there are no significant differences between abortion and infanticide – hence one cannot be justified without the other.

*Conclusion:*
- ➢ Be sure to link your conclusion back to the question by referencing the fundamental difference or similarities in ethical issues regarding abortion and infanticide, and why you feel the arguments you've favoured outweigh the arguments in favour of more/less testing.

*Should 'whistleblowers' be encouraged or discouraged?*

In this sort of question, 'whistleblowers' is, in part, a stand in for any subject – much of your essay should be spent arguing for whistleblowers to be encouraged, focusing on the benefits that whistle blowing allows. Alternately, you could argue that whistleblowers should be discouraged – in this case, you should make it clear why you believe this is the case and reject counterarguments, and cite examples to demonstrate your case (e.g.: you might argue whistleblowers can potentially expose severe problems and that despite the initial loss of confidence in the institution, the long term benefits are far greater). It is worth including several examples of where whistleblowers have helped or not, and therefore using these to support our argument.

You should aim to highlight the industries or areas of society where whistle blowing has been used in the past e.g. the NHS, athletics. You can mention how many people who notice wrongdoing often choose not to expose the truth, due to pressures from their peers and their seniors. Therefore, this question asks you to draw both on the ethical as well as the practical arguments for/against whistle blowing.

*Introduction:*

- Define what a "whistleblower" is and provide an example demonstrating your knowledge of whistle blowing.
- Clearly state your opinion and outline your main arguments.
- Link your answer into a broader understanding of the concepts mentioned in the question by stating why "whistleblowers" ought to be encouraged or discouraged, and the implications that might have on certain aspects of society.

*Why they should be encouraged:*
- The main reason whistleblowers should be encouraged is because the truth that they expose can be extremely important. For example, it could unveil a crisis with a particular industry or highlight an impending disaster. This reason is best supported with an example. An example which can be used is whistle blowing in NHS trusts such as the Mid-Staffordshire Trust scandal which exposed how several patients were treated with extreme negligence, and often left for hours unattended unable to go to the toilet. Whistle blowing allowed an investigation to take place so that patient care could be improved to adequate levels, thereby helping to improve patient care and survival rate in the future.

- Another example that could be used (especially in the recent climate) is that whistleblowers exposed cheats in sport. For example, due to whistle blowing there has been an investigation into state sponsored drug cheating in Russia which has caused Russian athletes to be banned from the Olympics, thereby hoping to help eradicate doping in sport.

- The next argument to be made in support of whistleblowers focuses on a more ethical aspect, and that is the freedom of speech. If people are aware of a problem, they should be allowed to highlight this problem to the general public so that it may be treated. As humans, we all have a right to say and do what we want as long as it is in accordance with the law – hence whistleblowers should be encouraged.
    - **Counterargument**: Although whistleblowers have a right to freedom of speech; we can look at the ethical debate from a different angle, which focuses more on the consequences of one's action. This theory of consequentialism states that an action should be deemed right or wrong by the effects it has. Hence, as whistle blowing can lead to a loss of confidence in the profession, which would be a disaster for some industries like the NHS, it should be avoided.
    - **Counter-counterargument**: On the other hand, we may counter this argument by considering duty based ethics. It is possible to argue that we all have a duty to do the right thing and expose the truth, and therefore morally, whistle blowing should be encouraged.

- It is possible to argue that whistle blowing is essential to help create a fully functioning democracy. In a democracy, all members have the right to know all the information so that they can come to an informed decision as to how they want to vote in the future. Withholding information form the public therefore undermines the current democratic system.

### *Why they should be discouraged:*

- One of the main reasons why whistleblowers should be discouraged is because they can cause a loss of trust in the profession. For example, the vast number of scandals concerning cheating and blood doping in sport has led to a vast decrease in the confidence of certain sports such as athletics. This problem has become so great that many athletes now live under the continuous suspicion of cheating even though they have been meticulously tested and are clean. This problem is especially serious in the healthcare profession, where the confidence is critical to establishing a trusting doctor-patient relationship. Anything, which weakens this relationship, such as whistleblowers, should be avoided.
    - **Counterargument**: One may consider the timeframe when assessing whether whistleblowers should be encouraged. Although in the short term it might lead to loss of confidence, over time the confidence in the profession may increase and patient safety standards increase.

- When employees join an organisation, they are required to sign a confidentiality agreement. This is important as it states that they cannot divulge sensitive information about the company or its employees to others, and so "whistleblowers" are effectively breaking the law.

- Whistleblowers often are sacked by their employers and may even face death threats from those whose information they leak. An example of this is Edward Snowden, who founded the website WikiLeaks – an online platform where it is possible for whistleblowers to share sensitive information. Wanted by the USA, he currently seeks political asylum in Russia. Hence, whistle-blowing can have a great negative effect on one's life and so should be discouraged.

### *Conclusion:*

Once again, engage with the question on a wider level by restating the ethical and practical reasons whistleblowers should be encouraged/discouraged and the implications this may have for certain sectors, such as the NHS.

*If inward migration is bad for a country, does it follow that outward migration would be good for that country?*

There is some ambiguity in this question, regarding what their terms "bad" and "good" mean. Most likely, they refer to the effect of society, which can further be categorised into economic and social factors. It is worth noting briefly in the introduction exactly how you interpret "migration" to avoid confusion, and to show you have thought that it represents both short and long term migration. With this type of question, we must accept that inward migration is bad for a country, rather than dispute the claim made in the question. Therefore it is worth noting the disadvantages of inward migration, and using these to either support or counter your proposal of why outward migration might be good/bad for that country.

One of the most important things to consider when answering this question is the exact wording of the question. It is not asking for the advantages and disadvantages of outward migration. Instead, the question asks you to consider the problems associated with inward migration, such as pressure on service, and then evaluate whether or not outward migration would help remediate this issue at all.

*Introduction:*
- Define your terms as discussed above.
- Accept that inward migration is bad for a country, and highlight some of the reasons why this may be the case, e.g. social, economic reasons.
- Clearly state your opinion and outline your arguments for it, as well as your reasons for dismissing counterarguments.

*Potential arguments why outward migration would be good:*
- One of the main problems with high levels of inward migration is potential overcrowding, which puts pressure on public services such as the NHS. Therefore, outward migration might help to reduce this strain on public services, thereby helping to improve patient care for example, but overall contributing to a better quality of life for the remaining citizens.
- Since inward migration can lead to depression of wages due to increased supply for limited demand, it is possible than outward migration may help to increase wages.
- There is an argument to be made that outward migration can be beneficial to a country in the long term. Although there is a loss of a working population, outward migration provides an opportunity for those emigrants to go to other countries, learn their trade and gain skills, which they can bring back to their original country.

*Potential arguments why outward migration would not necessarily be good:*
- One of the major disadvantages of inward migration is the larger influx of unskilled workers, which reduce wages for lower paid jobs. However, this does not necessarily mean that outward migration would be better, since it can result in highly skilled workers leaving the country in search of better prospects economically. This "brain drain" is prevalent in third world and developing countries, where talented individuals leave their country of origin to move to developed nations like the USA and so provides a great example of where outward migration is bad for that country.

- Outward migration can be bad for that country due to the fact that most emigrants are usually young, working class individuals. Therefore emigration often reduces the size of the workforce in the country of origin, thus causing the economy to shrink and with less tax collected by the government to spend on public services.

- Inward migration can often lead to overpopulation within a country. However, although outward migration may counteract this, the outward migration of young individuals can lead to an age imbalance in the country of origin, as the demographic is often skewed to a much older population. This has a great effect socially as it can lead to several towns losing their character and atmosphere whilst even leaving other areas completely deserted.

- Inward migration is often deemed "bad" socially as it can lead to increased tensions and even conflict between the native and arriving populations, eventually leading to segregation in society and local communities. There is nothing to suggest that outward migration does anything to resolve this problem, as regardless of who emigrates, the segregation between different communities still exists.

*Conclusion:*

Link your conclusion back to the question by stating to what extent outward migration may benefit a country, and discussing whether this outweighs the negatives of inward migration.

**END OF PAPER**

# 2016

## Section 1

**Question 1: D**

This passage starts off discussing the increase in rainfall over the past 50 years. It then moves on to suggest that India and China are prepared for heavy rainfall but this is not the case in the UK. It finally gives another reason for the increased rainfall. Therefore, the main argument of the passage is that the UK should do more to prepare for flooding and hence the answer is D.

Let us look at why the other options are wrong:

- A- This at first sight seems like a plausible answer but it is not the main argument. We can infer that the UK has been underprepared for flooding in the past but the main issue in the passage that the UK needs to be more prepared now because of the increased risk of flooding
- B-This is not the main *conclusion* of the passage. The conclusion is that the UK could learn from India by planning its infrastructure to deal with flooding
- C-There is no mention of political will in this passage
- E-There is no indication of this in the passage

**Question 2: C**

The quickest way to do this is to find some solutions and look for a common number. Working modulo 2, it is obvious that we need to select either 2 odd numbers and 1 even number or all 3 even numbers. This gives us the following valid combination (8,18,24) and (18,11,21). The answer is therefore **C**.

**Question 3: C**

Reading the passage, the sudden conclusion does not seem like a legitimate argument and we are asked to find the main flaw. With this question, it is sometimes worth spending 10 seconds trying to find the flaw without looking at the answer choices. If you can find a flaw without any hints, you are most likely to be right. We are told in the passage that there is a correlation between brain structure and sleeping and there is a link between brain structure and mental health. There is nothing in this passage that suggests changing one of these things will lead to a change in another and therefore the answer is **C**.

SECTION ONE                                                    2016

Let us look at why the other options are wrong:

- A- While this group is neglected, this group of people are not central to the argument so this cannot be the main flaw
- B- Completely unrelated to the argument
- D- The argument states that night owl should try this change 'if they wish' so there is no reliance on the night owls wanting to change their sleeping patterns
- E- This is actually a flaw of the argument but it is not the main flaw. If C is false, then the whole argument is false. If E is not true, then the argument is only partially flawed.

## Question 4: E

The "conclusion" style questions are very common in the TSA and you should practise these until you are confident in them. A popular method is to attach each option to the end of the paragraph and see if it makes sense. With this paragraph, the start is about the method of phasing out small denominations and it then goes on to say how such a move would hurt charities and consumers. The conclusion of this article is therefore **E**.

Let us look at why the other options are wrong:

- A-The paragraph suggests that charities benefit from small coin donations but nothing about whether people should donate to charities
- B-The paragraph only discusses small denomination coins so such a general conclusion cannot be true
- C-There is no suggestion of this in the paragraph
- D-Although the paragraph does suggest that those who save small coins would eventually benefit, it does not say that this course of action should be followed

## Question 5 :B

These questions are usually more subtle than other reading questions on the TSA and as such can be harder. It is often helpful to think of the underlying assumption when reading through the passage. The main argument of this passage is that when people lie, they have small micro-expressions which other people recognize and realize that the first person is lying. An assumption of this argument is that we are able to recognize micro-expressions so the answer is **B**.

In this case, it is also useful to arrive at the answer by eliminating the other answers:

# SECTION ONE          2016

- A- while the statement might be true, it is not an assumption of the main argument but a side case that should be considered
- C- this is an unusual one because it is an assumption of the text. However, the main argument of the text is that one shouldn't lie because people will recognise the micro-expressions. One of the examples of this is pretending to recognise someone so it is just an example of the argument and not an assumption of the core argument
- D- Just as the answer to C, the example of meeting someone in a supermarket is an example and assumptions do not lie in examples
- E- This assumption does not really make sense as the advice "do not lie" does not apply to someone who never lies. If an answer choice is confusing you, remember that this could be its purpose.

## Question 6: B
This is a wordy question that aims to confuse the reader so we must read the question carefully before attempting to answer. The first thing to do with time zone questions is to convert to one standard time zone; let this be GMT. Converting 06:15 local Auckland time to GMT give 18:15 GMT on 20 August, exactly 24 hours and 45 minutes after takeoff. Two and a half hours were spent on stops so total time flying was: 24.75-2.5=22.25 hours

## Question 7: C
This a reading comprehension question that is easy once all of the question is understood. We are simply required to find the row of the table that satisfies all of the requirements in the question. The maximum limit of $400 leaves Intimidator, Top Thrill Dragster, Voyage and Kingda Ka as the only possible options. The min height requirement of Intimidator and Top Thrill Dragster eliminate these two as Anton's height is only 143cm. This leaves Voyage and Kingda Ka as possible options. Of these, only Kingda Ka had a speed of greater than 90mph.

## Question 8: D
These 3D non-verbal reasoning questions are often the hardest of TSA section 1. To solve this question, we will look for certain conditions imposed by the two views and rule out certain nets based on these conditions. Looking at the first view, we can see that the top of the letter A is adjacent to the bottom of letter B. In cases C and E, the bottom of letter A lines up with the top of letter B so these two answers are wrong. Looking at the second view, it is clear that reading from letter E to the right gives letter A. In answer choice A, B is to the right of E and in answer choice B, an upside down A is to the right of E. The only correct answer is therefore D.

## Question 9: A

The main argument of the passage is that to increase blood donors, only those willing to give blood or those who have given blood should be eligible to receive blood. They argue that this is a fairer system of blood donations. This system is clearly not fair if there exist individuals unable to give blood so the answer is A.

Let us look at why the other answer choices are wrong:
**B.** The greater demand for certain blood types suggest that we should place more emphasis on getting blood donors from these types but does not weaken the fairness of the system in question
**C.** This suggests many blood donors have had transfusions but does not say anything about the number of blood transfusion patients who are donors
**D.** This statement fully supports the passage so cannot be the answer
**E.** This fact is irrelevant because the passage is about blood donations

## Question 10: C

When dealing with this kind of question, first try to summarise the steps of the argument. Then look at how each step leads on to the next and try to identify the flaw. The argument in question is that social media increases revenue. Since social media will become big in the future, it must be utilised to maximise profits. The key flaw in this question is that revenue is used interchangeably with profits. While social media use seems to increase revenue, it is wrong to assume that it increases profits. The answer is therefore C.

Let us look at why the other answer choices are wrong:
- A- the argument states that if a business uses social media, its profits increase. The argument is conditional on a business using social media so A cannot be right
- B- there is simply no such assumption made
- D- once again, the argument is conditional on a business using social media and does not say anything about a business using other forms of communication
- E- it assumes that social media use maximises profits but does not say anything about getting the maximal 18% growth rate

## SECTION ONE — 2016

**Question 11: B**
One of the best way to deal with conclusion questions is to see which answer choice would most naturally append to the paragraph in the question. The passage states that the apostrophe is of limited use because we can speak without indicating its presence. The passage also states that is decreasing in use in business and there is frequent misuse. All of this suggest that there will be no loss if the apostrophe is lost from the English language leading to answer choice B.

Let us look at why the other answer choices are wrong:
- A- Although the passage accepts that misuse of the apostrophe is rampant, it does not condone this behaviour
- C- The passage states that the apostrophe is becoming less frequent in business correspondence but does not say anything about its necessity in business correspondence
- D- This is definitely true and can be inferred from the passage. This is however not the main conclusion, it is only an example of the main conclusion
- E- The passage does not suggest that we should or should not use apostrophes- it only suggests that apostrophes have very little use

**Question 12: C**
This is another "riddle-style" problem that we must approach with a logical mind. If Tom works four nights without a consecutive period of more than 2 nights, he must be on shift for: Monday, Tuesday, Thursday and Friday. This means that Tom and Robert are on duty Friday night; Sheila therefore cannot be on duty on Friday. If Tom is off duty on Wednesday, bother Robert and Sheila must be on duty on Wednesday. The only possible answer choice satisfying these two criteria is C.

**Question 13: C**
This question clearly has a straightforward approach: calculate the cost for all possible combinations and choose the cheapest. This is going to be time intensive and we must find a quick way of doing it for the TSA. The younger sister will only be three years old and so gets free entry. This means that they can be completely ignored for the purposes of calculating the minimal cost. This leaves: one employed adult, one unemployed adult, one senior citizen, one senior citizen spectator and two children. The party is happening on a Saturday afternoon so this is clearly going to be during peak hours. With only two children, we can get one spectator ticket so we need to figure out how to get this ticket to minimize cost.

Spectators have the lowest ticket cost so it makes sense to pay 70p for a spectator ticket. We can then either get the mum an adult ticket (£3.80) and a concession family ticket (£4.80) or we can get the grandpa a senior citizen ticket (£1.90) and a family ticket (£9.50). These give a total cost of £9.30 and £12.10 respectively. The previous is therefore the minimal cost. This was quite arduous and not fully rigorous but a rigor is not needed for the TSA.

**Question 14: B**
There is nothing challenging about this question; we must simply look at the five answer choices in the question and identify if they are present in the pattern. Answer choices A and E are both present at the bottom of 'W'. Rotating answer choice C by 90 degrees gives the central tile of the letter 'S'. Rotating choice D by 90 degrees counter-clockwise is also a tile in the letter 'S' so the answer to the question must be B.

**Question 15: B**
This is a classic logic question where we must use only the information given in the passage. The main idea of the passage is that children with less regular sleeping schedules performed worse on a series of tests. It is therefore clear that having bedtimes at different hours would decrease cognitive development in children giving answer choice B.

Let us look at why the other answer choices are wrong:

- A- This statement suggests the opposite effect which has no grounds in formal logic and therefore this answer choice is wrong
- C- There is no mention of parents so this answer choice must be wrong
- D- There is no indication that the damage occurs throughout life so this answer choice is not correct
- E- There is no mention of cognitive disorders so this answer choice must be wrong

**Question 16: B**
The paragraph suggests that spending priorities in prisons are wrong as more money is spent on the internet and TV compared to the money spent on books. It then suggests to reduce reoffending rates, literacy rates must be improved. The assumptions of this passage are: increases in literacy rate correspond to decreases in reoffending and internet/TV have no literacy benefit. The latter corresponds to answer choice B so this is the answer.

# SECTION ONE  2016

Let us look at why the other answer choices are wrong:
- A- There is no indication that prison is easy for the inmates
- C- There is no mention in this passage of prisoners rights
- D- The passage makes no assumptions about intelligence of the prison population; it just suggests that no matter how clever prisoners are, increasing their literacy prevents reoffending
- E- The only wish of the author is to reduce reoffending rates- they do not express an opinion on the purpose of prisons

**Question 17: C**
The passage can be summarised as follows: religion does not provide a basis for moral framework. The Greeks considered how to live life and invented the word "ethics". However, Greek gods often misbehaved with some in ancient Greece being atheists. Therefore lack of religion does not imply lack of morality. Clearly, if the ancient Greeks had doubts in the purpose of morality, then this argument would be significantly weakened so the answer choice is C.

Let us look at why the other choices are wrong:
- A- This statement does not change the fact that some Greek gods were adulterous
- B- We do not have any information on Greek poetry or drama so the answer cannot be B
- D- This does not suggest anything to do with morality
- E- While this may be true, it does not indicate anything about the link between morality and religion

**Question 18: C**
This is a hard question to do in a rigorous manner but we must be careful to check all cases before coming to a conclusion. The easiest way to approach this is to start with answer A, see if it is possible and if not, move down the list. Note that the this network is configured such that waking two paths always has a greater distance than one path. Our path must therefore consist of four edges with each town being visited only once. The first answer is 78km and this can be achieved if the four minimal paths are taken (17,18,20 and 23). This cannot form a full path so we move onto 83km. This path is achieved by the following edges: 17,18,23,25. This is also not a full path so we move onto 86km. This can be formed by: 18,20,23,25. This is a full cycle and is therefore the answer.

## SECTION ONE     2016

**Question 19: A**
This is a very wordy comprehension question with a paragraph whose only purpose is to confuse the reader. If we keep the final goal in mind, the question becomes trivial. We are asked for the number of non-fatal serious injuries on non-built-up roads. We are given the number of serious/fatal injuries on non-built-up roads (1724) and the number of fatal injuries on non-built-up roads (187). The difference between these two numbers gives the required answer: 1724-187=1537

**Question 20: E**
This is another 3D non-verbal question that is hard to verbally explain. To do this question, we must view the structure from the direction indicated by the arrow and visualise the image. In this case, we would see 1 cube on the top row, 3 on the middle row and 4 on the bottom row. This leaves answer choices B or E. It is clear that the row of four extends the row of three to the left giving answer choice E.

**Question 21: A**
The argument of the passage is that farmers in Africa should be allowed to poach lions as Europeans killed bears. When stated in such simple terms, the flaw in this passage becomes clear: Europeans failing to protect their wildlife does not justify Africans failing to protect their wildlife. This is an example of a *Tu quoque* error- the argument that something is right is based on the fact that someone else does it.

Let us look at why the other choices are wrong:
- **B.** This statement provides evidence for the example given in the argument so it strengthens the argument, not weakens it
- **C.** This again strengthens the argument that African farmers should be allowed to kill lions
- **D.** This is a valid reason for lion conservation but does not address the argument in any way
- **E.** This is once again another valid reason for lion conservation but it fails to address the crux of the argument

**Question 22: B**
To see why the answer is B, we will show the equivalent statements from the question and answer:

1. The essay is of a far higher standard= My partner has given me flowers
2. The student is clever or the student copied= He must be feeling guilty or this must be a special occasion
3. The student is of average intellect= This is not a special occasion
4. The student copied= My partner is feeling guilty

# SECTION ONE 2016

Essentially, an occurrence implies two possible options. One of these options is wrong so it must be the other.

Let us look at why the other options are wrong:
- A- There is no statement detailing the two possible options
- C- There is no statement detailing the two possible options
- D- There is no conclusion after ruling out one of the two options
- E- Again, there is no conclusion mentioned after the two options are detailed

**Question 23: E**
The argument in the paragraph is that it is wrong for the companies to be paying their workers below a living wage and expecting the government to subsidise the difference. The principle behind this statement is that companies should be paying their employees enough to live on and the government should not be paying the difference. The main principle however relates to the living wage so the answer is therefore E.

Let us look at why the other answers are wrong:
- A- This statement is false and not supported by the text- it is the companies that are leaving holes in millions of paycheques
- B- The argument is about how the government has to assist people on low incomes because of companies
- C- This is true but simply restates part of the paragraph- it is not the underlying principle
- D- While this is true and in some sense relates to the underlying principle, it makes no mention of the companies which should be paying the living wage

**Question 24: B**
Let x be the number of incorrect answers in Round 1, y be the number of incorrect answers in Round 2 and z be the number of incorrect answers in Round 3. We can form the two following equations: $x + y + z = 9$ and $x + 2y + 5z = 22$

We also have that x is the largest of the three variables implying that x >3. Note that if z=2, there is no way to satisfy the second equation and the incorrect answers in round 1 or 2 cannot make the 22 points lost. Therefore z=3 and the equations simplify to: $x + y = 6$ and $x + 2y = 7$

This can be solved to give x=5 and y=1. The question asks for the value of x which is 5.

**Question 25: D**
The condition that the roll must have a width of 210mm rules out the first paper roll. The diameter must be less than 25.4mm ruling out the fourth paper roll. Buying in bulk is going to be more economical than buying a single roll but we cannot afford to buy the third roll in bulk. This means that the only remaining options are:

1. Buying the second roll in bulk- price/m= 13.14/125=10.5p/m
2. Buying the third roll as a single roll- price/m= 7.49/100= 7.49p/m
3. Buying the fifth roll in bulk= price/m=20.94/300= 6.98 p/m

The last option is therefore the most economical so we choose answer choice D

**Question 26: A**
The sizes of the patches are: 2,3,4,5 and 7. These patches are stitched together to form a rectangle whose height is twice its width. This means that the area of the patch will be a number of the form $2x^2$ where x is an integer. Note that removing an even number from the above set would produce four number with odd sum so cannot form the correct answer. This eliminates two answer choices. Now consider the sum when each odd number is removed:
♦ If 3 is removed, the sum is 18
♦ If 5 is removed, the sum is 16
♦ If 7 is removed the sum is 14

Only the first of these is a number of the form $2x^2$ and we therefore get the answer if ignore the piece with 3 squares (shape V). This short method without actually looking at any of the shapes will likely save time during the TSA.

**Question 27: C**
A lot of conclusion questions reveal the answer within the very first sentence. In this paragraph, we are told that there is no point to historical re-enactments before we are given more reasons for this to be the case. It is clear that the main conclusion from this passage is the pointlessness of historical battle giving answer choice C.

# SECTION ONE 2016

Let us look at why the other answer choices are wrong:
- A- While this statement is true, it is not the main conclusion; it is evidence that lends itself to the main conclusion
- B- This is not true as the passage states that there is a meticulous attention to detail
- D- This is true but there is no mention of this in the passage so it cannot be correct
- E- This is another point that leads on from some of the evidence in the passage

**Question 28: A**
The passage states that there is a correlation between levels of leads in the body and the violent crime even though the former precedes the latter by 20 years. It also states that lead has been banned in certain products in Europe for about 20 years so it is possible that there will be a fall in violent crime just about now (20 years after the 1990s). This leads to answer choice A.

Let us look at why the other answer choices are wrong:
- B- the passage only explain that lead is a cause of violent crime- it does not state that lead is the only cause
- C- once again, the passage does not say anything about lead being the only reason for violent crime so this answer cannot be the case
- D- while this may be true, there is no indication of this in the passage so cannot be true
- E- the passage states that the group in question are *more* likely to go on to commit violent crimes not that they are *highly* likely

**Question 29: C**
This passage is about the reasons for employees leaving a company. It cites discontent with superiors as the main reason. It then argues that people leave companies because they do not have conversations with their superiors about their expectation. It is suggested that staff turnover could be reduced if these honest conversations were had. The assumption in this statement is that people do not have these honest conversations so they are not honest when they are leaving their companies.

# SECTION ONE   2016

Let us look at why the other answer choices are wrong:
- A- It assumes that some expectations are realistic, not all of them
- B- The supervisors being difficult does not preclude workers from having honest conversations
- D- The argument states that if staff turnover is to be reduced, then honest conversations must be had- it does not assume that staff turnover should be reduced
- E- It actually assumes the opposite

### Question 30: A
Let the length of the margin be X. Then we can form a quadratic equation and solve for X:

$(24 - 2X)(18 - 2X) = \frac{18 \times 24}{2}$

$432 - 84X + 4X^2 = 216$

$X^2 - 21X + 54 = 0$

$(X - 18)(X - 3) = 0$

Since X=18cm is not a valid solution, we have that the margin is 3cm.

### Question 31: A
This question is a lot of text for a very simple answer. We are given a set of prices and are asked to find the cheapest option. The weekend prices are clearly more expensive so these options can be ignored. The box tickets come to £80 per person so are more expensive than the Balcony tickets. The box tickets can therefore be ignored. Finally, 85% of £95.50 is greater than £78.20 so these balcony seats are indeed the cheapest option.

### Question 32: D
I will refer to the graphs as 1-4 with graph 1 being on the left and graph 4 being on the right. It is clear that the regular structure of graph 1 lends itself to container C. Similarly, the two part linear structure of graph 3 lends itself to container E. Graph 2 has a linear portion followed by a portion with decreasing gradient. This suggests that there is a regular section followed by a section with increasing width giving answer A. Graph 4 suggests an initially widening portion followed by a narrowing portion and then a portion of uniform width. This suggests that glass vial in B leaving D as the answer.

# SECTION ONE    2016

**Question 33: A**
The paragraph suggests that the independence of constitutional courts and central banks is only in name as they are appointed by politicians. Furthermore, giving too much power to these unelected institutions would be undemocratic. The argument in the passage is that if a politician chooses the independent members, then these members are likely to be partisan going against the independence of these institutions. This argument would therefore be weakened if these appointees go against their nominators politically.

Let us look at why the other options are wrong:
- **B.** While this may be true, the passage is about the principle of independent officials and not the logistics
- **C.** A politician vouching for an independent candidate does not make the candidate seem any less partisan
- **D.** This supports the argument that independent candidates are likely to be involved in the political process
- **E.** This supports the argument that political appointees are likely to have vested interests

**Question 34: C**
The argument goes as follows: genius is either due to genes or environment. There are cases of 'genius' in very early life when environment can have little effect therefore genius is wholly due to genes. The flaw in this argument is that although environment seems to have little effect in one example, it is not possible to say that it has no effect.

Let us look at why the other options are wrong:
- A- Although there is an example with genius at an early age, this is not an assumption of the argument
- B- This might be true but it is not assumed by the passage
- D- This is a weakness of the argument style but not a flaw in the argument itself
- E- This is also a flaw of the argument but it is not a main flaw

## Question 35: D

The article explains that George Osborne's policy to remove housing benefit from council homes with spare rooms (bedroom tax) did not bring in sufficient savings. It then postulates that providing universal childcare would increase employment leading to more tax revenue. Therefore, the conclusion of this passage is that the government should implement a policy of universal healthcare. This question can be tricky as some of the other options look quite enticing.

Let us look at why the other answers are wrong:
- A- This statement is completely true but it is only a reason given in the passage and not the main conclusion
- B- The passage does not mention that the bedroom tax only affects a small number of people
- C- Similar reason as above
- E- This is a tricky choice and it is very tempting to choose this as the main conclusion. However, the passage talk about both hardship and government income. The conclusion must therefore relate to both of these results.

## Question 36: D

Let the price of a coffee be C, the price of tea be T and the price of the service charge be S. We have the following equations: $5C + S = 121$ and $4T + S = 82$.

There may be clever ways to solve these equations but we will start will the smallest option and work our way up.
- S=1 implies 4T=81 and this has no integer solutions
- S=2 implies 5C=119 and this has no integer solutions
- S=4 implies 5C=117 and this has no integer solutions
- S=6 implies 5C=115 and 4C=76. There are integer solutions to both of these equations so the answer is D

## Question 37: A

This question looks like a complicated mathematical exercise but the solution is mushrooms are growing at a certain rate. Let us try to identify the pattern:
- On day 1, the mushrooms form a 2x2 square
- On day 2, the mushrooms form a 4x4 square
- On day 3, the mushrooms form a 6x6 square

We see that on day n, the mushrooms form a 2nx2n square. We need to know when the mushrooms will take over a 100x100 square. This clearly occurs when n=50 which is 48 days from Wednesday (today is day 3).

# SECTION ONE                    2016

## Question 38: E
This question has a straightforward method: calculate the scores of each participant and identify the odd one out. The scores can be calculated as follows:
- Pip=28*54
- Eve=27*56
- Bob=24*63
- Viv=22*68
- Nan=21*72

Before you start trying to remember the day of year 4 when you were taught how to do long multiplication by hand, stare at those products for a bit. It should become clear that four of these are a multiple of 3 (found by summing the digits) while one is not. Viv's answer is equivalent to 1*2 modulo 3 while the other answers are 0 modulo 3. This must therefore be the odd one out.

This is an important takeaway for the TSA, you will very rarely need to engage in tedious computation; there is often a much simpler shortcut.

## Question 39: D
To summarise the paragraph: patients are not eating too well because the quality of hospital food is very low. This is leading to patients 'fading' away and their health suffering. The conclusion of the article is that hospital food should be made more appealing in order to increase patient's willingness to eat it.

With a conclusion question, the best strategy is often to see which answer choice can be most appropriately appended to the passage.

Let us look at why the other answers are wrong:
- A- This is a general point that is implied by the article but the conclusion is that these better systems involve better food
- B- There is no mention of nurses in the passage
- C- There is no mention that the food currently provided does not satisfy the dietary requirements of the patients
- E- This could be inferred from the passage but it is not a conclusion of the main argument

# SECTION ONE  2016

**Question 40: B**

This paragraph is about democracy as a whole and can be summarised as follows. Political legitimacy is given to the government by the people of a country voting in elections. Therefore, a higher electoral will correspond to a more legitimate political system. The idea behind this argument is that people give a political system legitimacy by voting but there is an implied assumption that people consent to the political system as a whole by voting. The answer is therefore B.

Let us look at why the other answers are wrong:
- A- The argument is that democratic political systems have a higher turnout so this cannot be an assumption
- C- This is just a very general point that is not actually addressed in the paragraph
- D- Political legitimacy and common interests are different topics and should not be confused
- E- Non-democratic systems are not mentioned in the paragraph

**Question 41: A**

This paragraph argues that high-street banks should not be able to sell financial products because their staff are not trained to explain them. We are therefore looking for evidence that banks are mis-selling financial products. This evidence comes in the form of answer choice A so this is the answer.

Let us look at why the other answer choices are wrong:
- B- This could be seen to support the argument that many customers are tricked into buying financial products. However, the argument clearly states the issues is with bank workers not understanding financial derivatives so this cannot be used as evidence.
- C- This actually weakens the arguments as it suggests that customers will have all of the necessary information when they make a decision
- D- This again weakens the argument as it suggests that customers can benefit without understanding the actual financial product
- E- This does not strictly relate to the point about workers not understanding financial products so cannot be used

**Question 42: C**
This question almost gives us too much information and we can solve two different equations to arrive at the same answer. Firstly, let X be the weekday price of potatoes in pence. Using the first statement that £3 would buy 3 kilos less potatoes on a Saturday morning, we can get the following equation:

$$\frac{300}{X} = \frac{300}{(X+5)} + 3$$

This is a simple quadratic equation that can be solved to give X=20 or −25. Since the price must be positive, we have X=20p/kg. Note the same answer can also be solved by forming an equation based on the second statement:

$$\frac{300}{X} = \frac{300}{(X-5)} - 5$$

This shows that solving quadratic equations quickly is a key life skill. In this case, both equations could be solved to confirm that you have arrived at the correct answer

**Question 43: A**
We have seen numerous examples of similar questions on the TSA and the first step is to rule out options that are not possible given the set of conditions. This leaves the following cars:

- Rover 820
- Renault Laguna
- Rover 825
- Ford Sierra

We then have to calculate the depreciation per mile for each of these. This is done by subtracting the mileage from 100000 to get the miles driven. The price loss is calculated by subtracting £1000 from the cost price and this is divided by the miles driven to arrive at the answer:

- Rover 820: (7000-1000)/(100000-30000)=6/70
- Renault Laguna: (7000-1000)/(10000-20000)=6/80
- Rover 825: (1500-1000)/(100000-90000)=1/20
- Ford Sierra: (2000-1000)/(100000-70000)=1/30

The lowest of these is the Ford so this is the answer.

## Question 44: D

With this kind of question, a systematic approach is needed to arrive at the correct answer. We will go through each statement and see if it is true:

A. The female numbers are higher than the male numbers above the ages of 34 so this is true
B. This is also true and can be seen by comparing the numbers in each age group
C. There are 417k men aged 85+ and 856k women aged 85+. This statement is therefore true
D. There are about 1.32 million in the 80-84 category while only 1.26 million in the 85+ age category. This statement is **not true**
E. This is true as it is the largest age group for both men and women

## Question 45: C

The main argument of this paragraph is that with the advent of technology, only young people are fully equipped to understand the challenges of the day. They are therefore uniquely poised to set the moral standard that we should live by and this is the conclusion of the passage. The answer is therefore C.

Let us look at why the other answer choices are wrong:

- A- Although this can be inferred by the passage, it is not the conclusion. This is the reason why only young people should set standards
- B- This is true and again is the reason that young people should set moral standards
- D- This is an example of how the older generation's standards are no longer valid
- E- This again is a reason for them to set the moral standards and not a conclusion of the passage

## Question 46: B

This is another logic question and we must clearly identify the formal logic that is being used in this statement. In this case, the "contrapositive" is used to arrive at a conclusion I.e. event A is predicated on event B, event B is not happening therefore event A is also not happening. We will show how this framework is shown by statement B:

- Rain only when clouds are present= solar eclipse when new moon is present
- No clouds therefore no rain= no new moon therefore no eclipse

Let us look at why the other options are wrong:

# SECTION ONE — 2016

- A- this argument is as follows: A depends on B. A is happening therefore B is happening (notice the difference)
- C- blood clotting is not used in the second statement
- D- this argument is as follows: A is a result of B. Therefore when B occurs, so will A
- E- this argument is as follows: A depends on B. B will happen at a certain time so A will also happen at this time

**Question 47: B**
This question illustrates that we must only use the question paragraph to answer the question no matter how ludicrous the actual answer sounds. To summarise the paragraph: prostitution should be legalised because it will always happen and the money spent on prosecution could be spent elsewhere. If prosecution is replaced with murder we get answer choice B so this is the answer, no matter how silly it sounds.

Let us look at why the other answer choices are wrong:
- A- This is contradictory to what is mentioned in the text so must be wrong
- C- The passage does not justify prostitution based on the personal circumstances of those involved
- D- The focus of police is not mentioned- we are only told that prostitution should not be the focus
- E- The passage makes no mention of rehabilitation

**Question 48: B**
The easiest way to do this is to find out the location of each moon at each of the 5 options:

|  | Othello | Hamley | Romeo |
| --- | --- | --- | --- |
| After 36 Days | $16/20^{th}$ of $2^{nd}$ orbit | $36/45^{th}$ of $1^{st}$ orbit | $36/120^{th}$ of $1^{st}$ orbit |
| After 72 Days | $12/20^{th}$ of $4^{th}$ orbit | $27/45^{th}$ of $2^{nd}$ orbit | $72/120^{th}$ of $1^{st}$ orbit |
| After 132 Days | $12/20^{th}$ of $7^{th}$ orbit | $42/45^{th}$ of $3^{rd}$ orbit | $12/120^{th}$ of $2^{nd}$ orbit |
| After 180 Days | $0/20^{th}$ of $10^{th}$ orbit | $0/45^{th}$ of $5^{th}$ orbit | $60/120^{th}$ of $2^{nd}$ orbit |
| After 216 Days | $16/20^{th}$ of $11^{th}$ orbit | $36/45^{th}$ of $5^{th}$ orbit | $96/120^{th}$ of $2^{nd}$ orbit |

Note that for every 72 days, the ratios all equal 0.6. Therefore, this is the only answer where all three moons are at the same phase and hence will appear collinear.

# SECTION ONE  2016

**Question 49: E**
We look at 12m van column and extract the following information:
- 4 hour van hire, 40 miles included, initial cost £47 and cost per mile extra is 30p
- 12 hours van hire, 120 miles included, initial cost £63 and cost per extra mile is 30p

It is cheaper to hire the van for the whole day when the extra mileage accounts for the difference in price between the vans. The difference in price is £16. Dividing this by 30p gets 53.3 extra miles so it cheaper to hire for the whole day when we need 40+53.3=93 miles of driving. The minimum whole number distance is therefore E.

**Question 50: C**
This is another numerical question that relies on the fact that we must work with whole numbers. In a 10m roll, a 2.5 section will contain 4 patters plus an extra portion of 10cm. As we need identical drops, we can only get 3 sections from each roll. There are 6.2m of wall width to cover with each drop being 50cm. This requires just over 12 drops so we actually require 13 drops. 4 rolls would get 12 drops but we need a final extra roll to cover the last bit of the wall giving us the answer 5.

**END OF SECTION**

**SECTION TWO**    **2016**

# Section 2

*Should we care more about the survival of animal species or the welfare of individual animals?*

This question concerns animal welfare and seems to ask what the best way of looking after animals is. There are two key things we must first clarify before trying to answer this question: why should we care about animals and who does this "we" refer to. In general, when a question uses the pronoun "we", it is referring to humanity as a whole and not a specific group of people.

Now we must consider why we should care about either the "survival of animal species" or the "welfare of individual animals". Here, there are many different ways of looking at the question and it is therefore wise to look at a few different reasons for caring. We can then decide which is the most important based on reasoned arguments. Finally, the TSA essay does not test specific knowledge; markers look for the coherence of your argument and general essay style. It is therefore not an issue if you do not have specific examples of animal conservation; your argument is much more important.

*Potential reasons for caring about animals:*
- They provide a useful source of food
- They are an important part of the ecosystem (e.g. bees pollinate plants)
- It is our moral duty to look after all life-forms

*Introduction:*
- Define the key terms of the question; in this case, we should define: survival and welfare
- List some of the potential reasons for caring
- Come down on a side of the argument

*Arguments for caring about the survival of animal species:*
- Caring about the survival of the whole species will have a greater impact- it is all very well to care about the treatment of each individual animal but such a focussed policy will not be as effective as one looking at a species as a whole. As an example, we can take the survival of bees, whose population has been decreasing in recent years. When there is such a large population, there is no gain in looking after the welfare of each individual bee; there will only be an effect if we look at the population as a whole. An example of such a policy

would be banning herbicides that are harmful to bees. Such a policy is important because bees are vital for the pollination of much of our ecosystem. This point shows that we do not need a strong example or even quote an endangered species; we need to explain a relevant point with good justification. A counter-argument to this would be the first point on the next section; by thinking about counter arguments, you show to the examiner that you can structure an essay properly.
- Approaching this question from a very primitive perspective, it is possible to argue that we should only care about animals because they are a useful source of protein and other nutrients. We therefore should not care about the welfare of individual animals, we only require there to be a stable population of animals to feed our needs. This is clearly a controversial point which again is fine for the TSA; it just needs to be explained clearly and carefully. When making this point, it must be explained why animal welfare does not concern us e.g. humans are at the top of the food chain for a reason

## *Arguments for caring about the welfare of individual animals:*

- It is only by caring for the welfare of each individual animal that we can ensure the survival of a species. There are certain protected species such as giant pandas whose populations have been threatened by poachers. The only effective way of preventing this from happening was to introduce strict punishments for those involved in poaching. Methods such a breeding in captivity would be fruitless if poaching were allowed to go on unpunished.
- We can also make a broad ethical point about our moral duty to protect life on earth e.g. it could be suggested that as a civilised society the welfare of all animals concerns us all. The overall survival of a species is irrelevant if there are animals being mistreated. This point here does not really need an example- we are simply making a broadly philosophical point.

## *Conclusion*

This is arguably the most important part of a TSA essay as we must demonstrate to the examiner that we have clearly answered the question. This is easy to do if our previous two paragraphs have addressed the different sides of the question. We must then explain which priority is more important and arrive at a conclusion e.g. on balance, it is clear that the welfare of individual animals is more important. By ensuring the welfare of individual animals, we ensure the welfare and therefore population of animals as a whole.

# SECTION TWO — 2016

*Do people vote in line with their personal economic interests?*

This question is fairly straightforward but we must consider what "personal economic interests" really mean. In order to understand this phrase fully, we will consider the term "personal" and the term "economic" separately. Please note that you are not expected to know the definitions of any technical vocabulary but you must be able to make a reasoned guess at any words you do not fully know.

The word "personal" can either refer to oneself or a close group including oneself. We will take the definition to be the latter in this case as it allows for more argument. For this question, we will define "economic interests" to be related to the net-worth of an individual. This question can now be read as do people vote for the party that would increase their net wealth. Finally, the TSA essay does not test specific knowledge; markers look for the coherence of your argument and general essay style. It is therefore not an issue if you do not have specific examples of animal conservation; your argument is much more important.

**Introduction:**
- It is important to remember that the introduction should be structured in three key parts: introduce the question, introduce your points and introduce your final argument- this means that you need to plan your whole essay before writing your introduction
- Introduce the question: define what is meant by all of the key terms in the question as we have done above
- Introduce your points: we will discuss some points below, these should be included here
- Introduce your argument e.g. "it is clear that economic interest do play a large part in how people vote"

**Arguments:**
- If someone's aim in life is to get more money, then the optimal strategy would involve voting for the party that gives you the most money. In the UK, there is a clear link between income and the chance of voting labour/tory. Labour is the more "left-wing" party in the UK and they have a higher proportion of low income voters. This is because labour traditionally have policies supporting those on lower incomes by taxing those with slightly more money. With this question, it is very easy to get stuck in a very detailed example but the point of the TSA is not to discuss one topic in a lot of depth. In this case, we have made the point of labour/conservative in the UK and we must analyse the principle: income bracket is correlated with how people vote in the political spectrum.

- Counter argument for above point- we need to consider the fact that other demographics may be a confounding variable. In the case US, it can be shown that individuals with higher levels of formal education generally vote Democrat (left-leaning party). Once again, it is important not to spend the whole essay discussing American politics. We have made our point regarding education linked to wealth and we must now explain why it is a counter-argument. If education level is the true driving factor behind how people vote, then they are not basing their vote on their income. This means that their vote might not align with their own personal economic interests
- People vote in alignment with their political beliefs and not their own economic interests- a feature of a more right wing government is that they believe in less state intervention (no "nanny state") and that individual should be able to get on with their own lives- this can be seen most clearly with the Republican party in the US who believe in small federal government. Among certain agricultural workers, there is huge support for the Republican party as these workers do not want the state meddling in their affairs. However, such workers are already on quite low incomes so may actually financially benefit from increased federal spending. In these areas of the country, there is a certain pride with voting Republican even if it is not in their best interests. Here, we have once again given a specific example but we must extract the broad principle to form a coherent argument: political philosophy and self pride are more important than economic self interest when determining who to vote for

**Conclusion**

This is arguably the most important part of a TSA essay as we must demonstrate to the examiner that we have clearly answered the question. This is easy to do if our previous two paragraphs have addressed the different sides of the question. We must then explain which priority is more important and arrive at a conclusion. As the conclusion is more weighing up both sides of the argument it is important to not get tied down in specific examples and compare larger themes. In this case, we could argue that there are two main reasons for people's voting patterns: their philosophy and the benefit they get from that party. The balance between these two factors is determined by an individual's circumstance so we cannot really come down on either side of the argument.

*Is it possible to over-regulate the banking system?*

This is a TSA essay that will be more suited to economist students but it should be accessible to anyone with the required interest in current affairs. This question is particularly relevant now as the Trump administration is weighing up whether to repeal the Dodd-Frank act. This question will admittedly be hard to answer if are not familiar with certain aspects of the banking system but there are still three other accessible questions in this case.

As always, it is important to define the key terms of the question in the introduction. In this question, the two key terms are "banking system" and "over-regulation". Although these terms seem fairly obvious, it is always good practice to provide a clear definition. This will allow the examiner to follow your line of argument, even if you slightly misinterpret one of the terms in the question. We define a bank as an "establishment that takes deposits from customers and invests its returning it to customers when requested along with interest". The "banking system" is a collection of retail, investment and other types of bank. Regulation can be defined as legal restrictions placed on the banking system and we will define over-regulation as excessive regulation

### Introduction:

- Introduction should be tripartite: the question should first be introduced and key terms should be defined (as mentioned above); the points of the essay should be introduced and then you should hint at your conclusion
- As your introduction needs to address the points that you will make in your essay, you will need to plan your whole essay before writing the introduction
- With the technical nature of this question, it might be worthwhile to introduce some of the key pieces of legislation

### Key pieces of legislation:

- Glass –Steagall act (1932) and Banking Acts of (1933)- These were introduced by FDR in response to the Great Depression and aimed to separate the retail and investment arms of banks. Although the act has a lot more technical content, knowing the broad principles is more than enough here.
- Gramm-Leach-Bliley Act- this act is not as well know but was passed into law by the Clinton administration of the 1990s- this essentially repealed some of the key statutes of the Glass-Steagall act and arguable led to the spread of the financial crisis.

- Once again, it is not necessary to know the details of the act or even the name but it is certainly worth knowing the deregulation of the banks may have precipitated in the financial crisis.
- Dodd-Frank act- signed into law by Barack Obama in response to the financial crisis of 2008- this was named the "modern Glass-Steagall" and once again separated the investment and retails arms of banks

**Arguments:**

- Over-regulation is most definitely possible and this could have a negative effect on the banking industry- one example of this is the "bankers-bonus tax" proposed by Ed Miliband in 2015 or even the gender quotas in boardrooms suggested by some on the left. Regulation could also relate to the operations of the bank itself. It is now necessary to explain why these examples are "excessive" I.e. we must show that these negatively impact the banking industry. One way to explain this is that with the international nature of banks, this kind of regulation would force banks to relocate to other countries with less regulations which would ultimately harm consumers. Furthermore, the separation of investment and retail arms would reduce returns to investors.
- On the other hand, we could argue that the banking industry cannot regulate itself because those at the top are motivated by money. These people therefore cannot be trusted to make decisions about the safety of deposits so government regulation is needed. To explain this, we could argue that bankers can take risks with their investor's money without being personally liable and this breeds a dangerous mindset.
- Finally, in this question it is necessary to consider a financial crisis and two seem most relevant: the Great Depression and the Financial Crisis of 2008- both of these led to loss of credit availability severely harming the whole economy- in fact, it could be argued that deregulating the banks led to the 2008 crisis so there is no such thing as overregulation.

**Conclusion:**

- Although you may have your own personal beliefs, for the TSA essay, it may be best to choose the side with easiest line of argument
- Here, it is easy to argue the over-regulation is not possible because the risks of too little regulation are far too great
- When there has been deregulation, a crisis has followed therefore any regulation reduces this risk and is therefore a worthwhile piece of legislation

# SECTION TWO — 2016

**With easy access to vast amounts of information through the Internet, what advantage is there to remembering facts?**

This is a very general essay question so could be answered if you have no specific knowledge needed for some of the other questions. It is also a very broad question that has a wide range of possible answers. As the question is so broad, it is important not to list lots of examples but have substantial points and then use examples to provide evidence for these points.

Although it is usually necessary to define all the key terms of the question, since this essay title is very self explanatory, it is not needed. If you however want to focus on a specific aspect of the topic, then it is definitely worth stating this clearly in the introduction.

### Introduction:
- It is important to remember that the introduction should be structured in three key parts: introduce the question, introduce your points and introduce your final argument- this means that you need to plan your whole essay before writing your introduction.
- Introduce the question: usually we would define the terms of the question but as this is not necessary for this question we could start off with a controversial statement: "Some people say that knowledge is not as useful as being able to use Google search"
- Introduce your points: we will discuss some points below, these should be included here.
- Introduce your argument e.g. nothing matches the speed, flexibility and robustness of the human brain.

### Arguments:
- Speed of recall- if one already knows a fact, then a significant amount of time is saved as the person no longer has to go on the internet, search for the correct webpage and find the answer. This time could be vital in the case of an emergency e.g. a random bystander might witness someone faint and decide to put them in the recovery position- if they already know how to do this, they will save time compared to looking this up on Google and completing the task potentially saving someone's life- although this is quite an extreme example, it is simple and illustrates the point; this is sufficient for the TSA

- Robustness of recall- the term robustness is essentially a measure of how a system reacts to small perturbations from normal operating conditions. This is quite a general description but it definitely the case that the human brain is much more robust than a computer at fact recall e.g. if there is a power cut, or problem with the internet, or the person searches for the wrong item on the internet, the computer will not be able to tell the user the correct piece of information. The internet relies on the correct information being supplied exactly and is therefore not robust. A robust system can deal with changing environments and can best adapt to current conditions- this could arguably provide an evolutionary advantage
    - This point is quite broad and could be framed as a counter argument to the amount of knowledge that a computer can store. Using a counter argument approach is definitely a good thing to for this question
- Interconnected nature of the brain- the brain is a complicated network with many billions of neurons. This means that the brain is uniquely able to gather information and group facts together. This may allow new facts to be discovered or it could allow for the consolidation of facts. For example, it may be possible to memorise the locations of all the stations on a train line but the brain has a unique ability to connect these stations together. This could provide new information on the route itself. Note that this is not a concrete example but such is the nature of the question- do not be afraid to adapt your essay style to best suit the question.
- The question specifically asks for advantages of memorising facts so only one side of the argument needs to be presented- it is however wise to consider counterarguments and dismiss them e.g. it could be argued that the internet is more accurate but we pay a cost in robustness. Robustness is arguably more important because it allows us to work in very different environments

**Conclusion**
- It is first worth summarising the broad nature of the points e.g. there are three clear advantages of learning facts: speed, robustness and interconnected nature- this shows the examiner that you have thought about your essay and not just written down a list of facts.
- To come to a conclusion, it is not necessary to continue with your argument from the main body- in fact, it is particularly easy to argue that as the internet grows in size it get the knowledge of more and more humans- this is clearly better than the knowledge of one human so memorising large amounts of facts is a fruitless exercise.

**END OF SECTION**

## 2017

# Section 1

**Question 1: E**
The key sentence here is that:
"The label "mad" is cynically used as a way of dehumanising and discrediting leaders of countries with whom we are in dispute." None of the other answers capture that the passage is defending those normally characterised as "mad"

**Question 2: A**
There are 3 stages to this question. The first is the sale of pizzas and flapjacks before the end of lunchtime, the second is the sale of pizzas and flapjacks during the afternoon when the prices is halved and the third is the remaining pizzas and flapjacks being given away for free.
In stage 1, 40 mini pizzas are sold (as 10 remain) and 35 flapjacks are sold (as 15 remain) at a price of \$2 and \$1: 40x2 + 35x1 = 115
In stage 2, 8 mini pizzas are sold (2 remain) and 14 flapjacks are sold (1 remains) at a price of \$1 and \$0.5: 8x1 + 14x0.5 = 15
In stage 3 the remainder are given away for free

Therefore final addition: 115+15 = \$130

**Question 3: E**
The concern raised in the passage is that as time goes on not enough people will vote to be able to bestow a mandate upon the government. This problem is solved if, as time goes on, the people who are currently not voting begin to do so with age

**Question 4: B**
The passage firstly states that the media environment is one of celebrity influence, then goes on to state that celebrities are role models and that to deny this is "unrealistic and irresponsible" implying that they ought to act responsibly due to their being inevitably role models

# SECTION ONE 2017

## Question 5: B
All of the concerns raised in the passage arise from the importance of appointing the best person to the job of being head of the company. If this were not important, then the fact that the sons of owners are less good candidates in general would become irrelevant.

## Question 6: D
In order to calculate the total height we need to calculate the height of the picture, the added height of the mount at the top and bottom of the picture and the added height of the frame at the top and bottom of the picture

Height of picture: 40cm
Height added by mount: 6cm (top) + 9cm (bottom) = 15cm
Height added by frame: 2cm (top) + 2cm (bottom) = 4cm

**Total height** = 40 + 15 + 4 = 59cm

## Question 7: A

Difference compared to the previous month:

|      | Jan | Feb | Mar | Apr | May | Jun | Jul | Aug | Sep | Oct | Nov | Dec |
|------|-----|-----|-----|-----|-----|-----|-----|-----|-----|-----|-----|-----|
| 2013 | -5  | -17 | -19 | 27  | -24 | 16  | 18  | -11 | 18  | -54 | 15  | -17 |

Largest **increase in 2013**: 27 – Apr 2013

## Question 8: A
Total area of garden: 25x10 = 250m$^2$
Total area of patio: 2x3 = 6m$^2$
Total mowed area = 250-6 = 244m$^2$
Time taken to mow with old mower = 244 ÷ 1 = 244 minutes
Time taken to mow with new mower = 244 ÷ 2 = 122 minutes
Time saved = 244-122 = 122 minutes

## Question 9: A

The passage states that few people are unhappy with their bank service because they do not have a tendency to change banks, which would be expected if they were unhappy and wanted to find a better alternative. The low statistic of 11% means that examining the large proportion of people who have changed banks and the significant proportion who have changed repeatedly significantly weakens the above argument as it seems to be the case that this 11% are anomalous and have found a suitable bank for them in childhood.

## Question 10: C

The passage claims that the initiative will increase awareness of poetry, which is likely true, but then asserts that this will improve appreciation of poetry. There is no justification provided for this step in the reasoning and it is not necessarily true if children simply learn poems robotically in order to win the competition

## Question 11: D

The passage focuses on responses to the increasing popularity to health clubs and therefore this must be present in the answer. The answer must also reflect the negative conclusion reached by the passage centred on self-deception.

# SECTION ONE 2017

**Question 12: C**
Using algebra:
Number of 19p coupons: X
Number of 12p coupons: Y
Number of 7p coupons: Z

Where X+Y+Z = 7
19xX + 12xY +7xZ = 100

Since there are at least one of each then X, Y and Z are at least 1

Therefore, we already have at least 19 + 12 + 7 pence = 38p

To find the remaining coupons:

19xX + 12xY + 7xZ = 62p
Where X+Y+Z = 4

We can't use any Zs as the total that we need from 4 coins is too high for them to be useful

Only possible way to get 62 from Ys and Xs is with 2 of each (19x2 = 38, 12x2 = 24, 24+38 = 62)

Totals: Xs (19p) = 3, Ys (12p) = 3, Zs(7p) = 1

**Question 13: D**
Follow each row to see which teams have no games in which they scored 0 goals at home (first number). This leaves us with Amazon, Ganges, Nile, Rhine and Volga. Check these columns only for which team did not score 0 goals in any game as the away team (second number). This leaves us with Rhine.

**Question 14: A**

Initial shape looks like this, the shapes
1, 2 and 5 fit together as shown to emulate
This 'L' shape

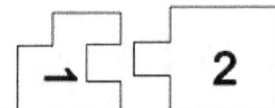

**Question 15: E**
The passage initially raises single use plastic bags but does not focus on them throughout, the change in focus to the production methods of alternative bags makes this relevant and means that it must be included in a summation of the conclusion as the single use plastic bag discussion was only a set up to this conclusion.

**Question 16: A**
The Passage focuses entirely on the issue of congestion and evaluates the policy only relative to its effect on congestion. If this were not relevant, the argument would suffer.

**Question 17: C**
The possibility of posting information anonymously adds a new dimension to the issue of censorship mentioned in the passage. This might, for example, reduce the ability of the government to censor individuals through the threat of prosecution if that prosecution could not happen in practice due to anonymity.

**Question 18: A**
Total surface area of walls and ceiling =
8x4 = 32 (ceiling)
(4x3) x2 = 24 (2 short walls)
(8x3) x2 = 48 (2 long walls)
Total area = 32+48+24 = 104m$^2$

Area to paint = 104-10 = 94

Area possible to paint with one tin: 8x12 = 96

Tins of paint needed = 94/96 = <1 = rounded up to 1

# SECTION ONE 2017

**Question 19: C**
Note that the number of points per improvement in performance increase as the performance gets better.

Test each of the possible answers in turn by calculating the difference in score between that score and the score for jumping 18cm shorter.

Start in the middle and if the difference is too small then try the higher jumps, if it is too large then test the smaller jumps.

Testing the middle score gives us 903 points (1.74m) – 689 points (1.56m) = 214 points

Answer found!

**Question 20: D**
Difference:

| Jan | 37.1 |
|-----|------|
| Feb | 10.1 |
| Mar | 77.4 |
| Apr | 29.9 |
| May | 25.8 |
| Jun | 35.1 |
| Jul | 33.6 |
| Aug | 18.7 |
| Sept | 60.1 |
| Oct | 18.3 |
| Nov | 10.2 |
| Dec | 16.7 |

January is approximately half of March – eliminating A, B, C and E

**Question 21: B**
As is a common fault in argumentation, this passage attempts to generalise among all unlicensed moneylenders. Since the conclusion of the passage can only be made if the moneylenders are indeed harmful in general, it is necessary to not just extrapolate a very small amount of data more generally

# SECTION ONE 2017

## Question 22: C
The passage presents us with two possibilities and then very explicitly eliminates one of them to leave us with only one choice. While B appears to do the same, it is not quite as definitive in its elimination of one of the two options.

## Question 23: B
The passage raises the points that there is a risk in everything and that there is also such a thing as personal responsibility meaning that an individual need not be controlled by the state unduly and should be allowed to be responsible for their own choices. These arguments are also applicable to the legalisation of drugs since the risk of drugs can also be defended by claiming that an individual has personal responsibility over their own body despite the risk of drugs.

## Question 24: B
Total mass of chemical: 6.0kg
Total mass of X in the mixture = $6 \times \frac{1}{4} = 1.5$
Total mass of Y in the mixture = $6 \times \frac{3}{4} = 4.5$

In order to reach a 40% - 60% ratio, it is necessary that there is 2/3 the quantity of X in the mixture than there is of Y as $40 \div 60 = 2/3$. Therefore, 3kg of X are needed in total so 1.5kg needs to be added.

## Question 25: E

|  | 1st choice | 2nd choice | 3rd choice |
|---|---|---|---|
| Jo | Portugal | France | Greece |
| Mel | Portugal | France | Tenerife |
| Kim | Majorca | France | Tenerife |
| Lexy | Greece | Tenerife | France |
| Naz | Tenerife | France | Majorca |

**Totals:**
Portugal: 6
Majorca: 4
Greece: 4
Tenerife: 7
France: 9 (**BUT NO FIRST CHOICES**) – therefore the answer is Tenerife

# SECTION ONE — 2017

**Question 26: C**

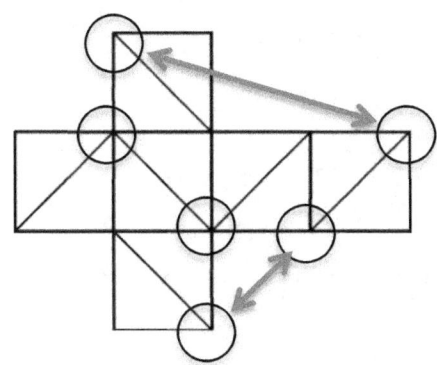

Relevant corners indicated

**Question 27: A**
The passage makes reference both to the fact that the HRA is important to those soldiers not in battle but also that it ought not to apply when soldiers are in battle. Both of these elements are separate and clearly relevant and so need to be included in the answer.

**Question 28: D**
The only answer that is explicitly mentioned in the text is D where it is said that: "the aquaculture sector has yet to devise effective methods for rearing some commercially popular predator species, such as tuna, without having to feed them wild-caught fish, which contributes to the pressure on our oceans" which is relevant since the point of aquaculture is to reduce pressure on the oceans.

**Question 29: E**
The experiment measured in the passage purports to measure the effect of chocolate on Alzheimer's disease but does not actually clearly measure any symptoms of the disease itself, only those things tangentially related to the disease.

## SECTION ONE — 2017

**Question 30: C**
Solve using simultaneous equations:
X = number of majors scored
Y = number of minors scored

For the Red team: 5X+3Y = 77 where 2X =Y
Therefore, 5X + 6X = 77 so X=7

For the Blue team 5X+3Y = 52 where 2Y = X
Therefore, 10Y +3Y = 52 so Y = 4 and X=8
**Total Majors = 7+8 = 15**

**Question 31: D**
Using fat or calories (due to their larger number of significant figures)
Amount of fat per oatcake: 2.2g
Amount of fat per 100g: 17.6
Number of oatcakes per 100g: 17.6÷2.2 = 8
Number of oatcakes per 300g packet = 8x3 = 24

**Question 32: E**

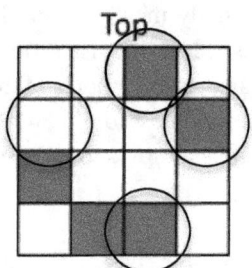

From examining the other 4 answers, they all have these four circled squares in common, meaning that they must form part of the correct solution. This solution does not have the top left square coloured meaning that one of the other two coloured squares must also be part of the correct solution as only one mistake could have been made. The answer is therefore E as it is the only solution to colour the four circled squares and one of the two squares in this answer that are not circled.

## SECTION ONE  2017

**Question 33: C**
In summation, the passage is claiming that the parents of more than two children are being irresponsible, this is further true if parents are having children that they did not intend or wish to be born when these children go on to have significant effects on the environment.

**Question 34: B**
The stated conclusion: that television dampens people's ability to think for themselves, does not necessarily follow from the fact that television causes an individual to be entertained rather than to think and discuss. This assertion is therefore not relevant to the previous claims made in the passage.

**Question 35: E**
This passage states its main aim initially by saying that we ought to change science lessons in order to give a correct view of modern science and then all other arguments after this point are made in support of that initial conclusion. The closing line that the traditional view is giving children a false, unscientific, view of the universe, is supporting evidence that the science lessons ought to be changed.

**Question 36: C**
In total, the boat takes 45 minutes from setting off before it is ready to set off from the other end. Therefore, the boat setting off from the mainland at 5 past will be ready to depart back when the boat is next scheduled at 55 minutes past from the island.

None of the other 4 times before 9.55am can be made by another boat travelling from the other destination so at least 5 boats are needed in total for the first journeys.

All journeys from one side have another journey leaving from the other side 50 minutes after that journey therefore there is **no need for other boats** as the turnaround for each boat is 45 minutes so the same boat can fulfil one journey and then the next 50 minutes afterwards = 5 boats.

## Question 37: D
Approximate the ratios of each area:

| England N | 20:1 |
|---|---|
| England Mid | 20:1 |
| England E | 18:1 |
| England LDN | 20:1 |
| England S | 18:1 |
| Wales | 20:1 |
| Scotland | 15:1 |
| N Ireland | 17:1 |

Scotland is the clear answer

## Question 38: C
Two identical pieces which when they fit together do not form a complete circle:

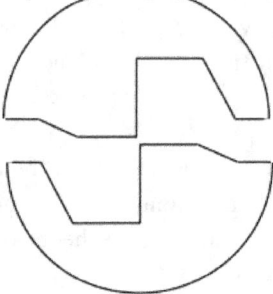

## Question 39: A
The passage provides significant evidence that there are many benefits to assigning individual caseworkers to the most vulnerable homeless people. All of the other answers are slightly too tenuous to be clearly drawn from the passage while the benefits of this pioneering scheme are obvious and therefore this conclusion can be clearly drawn.

## Question 40: E
There is a lot of information in this passage but in summary it focuses on how to best promote health attitudes among drinkers: with simple instructions or more complex ones. This would not happen if the complicated guidelines were in fact not read and understood at all.

## Question 41: A
This passage makes the very broad statement that regardless of the cause of an extinction level event; the planet would be able to recover swiftly. This is a very difficult claim to be able to make without knowledge of exactly what would happen in the instance of every possible extinction-level event. A global nuclear holocaust for example has never been witnessed and therefore may have an effect that this argument has failed to account for.

## Question 42: B
Cost each day from local pet shop:
2 sachets at $1 = $2
25g at $1 per 100g = 25p
Total = $2.25

Cost each day from online distributor:
One sachet = $62.40÷96 = approximately 66p
1kg of dry food = $8, 25g of dry food = $8÷40 = 20p
2 sachets = 1.32
25g of dry food – 20p
Total = $1.52

Difference is approximately 75p

## Question 43: A
Only Arps and Urps have tails and they have one each, since there are 33 tails there must be a total of 33 Arps and Urps. Therefore the remaining 12 animals are Orps.

Subtracting the total 48 legs and 24 horns from the 12 Orps from the total leaves 174 legs and 75 horns. Since only the Arps have horns out of the Arps and Urps, the remaining 75 horns must come from the Arps. Therefore there are 75/3 Arps = 25 Arps.

Therefore the remaining 8 animals must be Urps

**Question 44: A**
The sections on the net in A circled (below left) do not cover the side on the completed shape circled (below right)

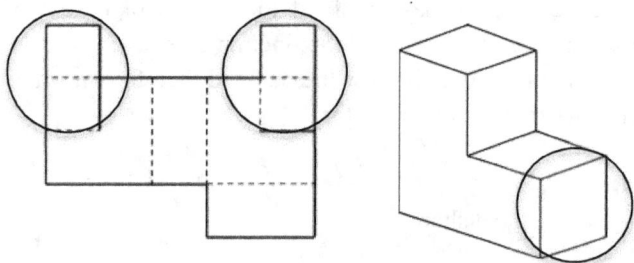

**Question 45: C**
The passage focuses on the social effects of the Buy-to-let policy on both oneself and the wider public. The unaffordability of houses and the lack of individual benefit due to the increase in rent far above wages, and the risk of investment, has shown that there is no social benefit available to anyone

**Question 46: B**
This passage claims that a particular characteristic inevitably implies the absence of two other things; in this case the freedoms of speech and assembly. This parallels the idea that a constant amount of energy in the body implies a lack of light and heat being emitted.

**Question 47: B**

The passage states that in a given situation it is always possible to act for the greatest good for the greatest number and also clearly states that this is what we ought to do. 'Ought' then implies that the action which benefits the most persons is the right action and therefore is the best possible action.

## SECTION ONE 2017

**Question 48: D**
The difference between RAT and TEAR is only the letter E; therefore the difference in points between the words is the value of the letter **E**.

Now that we know the value of the letter E, we subtract 2 times this amount from TREE to find the value of TR. The difference between the value of TR and RAT is then the value of the letter **A**.

Now that we know the value of TR and E, we subtract this from RITE to find the value of **I**.

No further letters can be found

**Question 49: D**
Start with the earliest time zone (San Diego) at the earliest available time (8:00). At this time Barcelona (the second latest time zone) is at 17:00 and therefore will be available for **3** more hours.

When London (second Earliest time zone) is then at 8:00, Nairobi (latest time zone) is at 10:00 and available for 10 more hours.

These 2 periods crossed over for 2 hours, totalling 11 hours.

**Question 50: D**
It is not possible to rotate the tables on the left to create the shape on the right

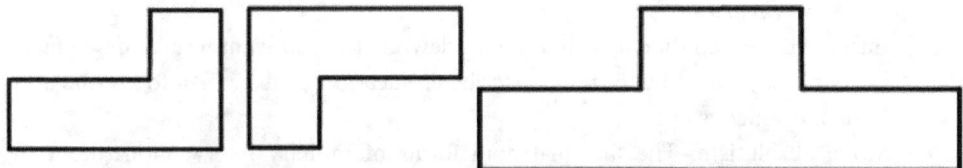

## Section 2

**Are 'drone strikes' morally different to military campaigns fought on the ground?**

*Introduction*
***ADVICE:***
- This is a complex question that can be approached from a significant number of directions, and which will produce a variety of valid answers depending upon which ethical viewpoint one adopts. I would stress that, given the limited amount of space available to TSA candidates for their essay, they ought to think very carefully about whether it is worth spending all of their time attempting to establish a case for a particular ethical view when they have so little space within which to do it.
- One approach to this question would be to consider a variety of ethical viewpoints and how the practicalities of this policy might affect them, in order to make this answer approachable and easily comprehensible I would recommend the following approach:
    - Consider this policy in terms of moral absolutes i.e. 'killing is wrong' since both clearly do kill, and both also clearly breach the sovereignty of a nation
    - Consider this policy in terms of Utilitarianism which is likely to claim drone strikes to be ethically superior

***WHAT TO INCLUDE:***
- Include a very brief description of your main points
- Set out what your conclusion will be

***Philosophical theory***
- Ensure you are confident with a moral viewpoint if you want to include them in your essay – it is better not to ascribe an idea to a thinker than to ascribe a false or erroneous one.
- Moral absolutism: The fact that both forms of military intervention (drone strikes and 'boots on the ground' invasions both lead to the outcome of death of enemy combatants means that both might be regarded as morally indistinguishable by a moral absolutist who believes that killing is wrong. This is also true for those who uphold the sovereignty of nations as an absolute. Since the sovereignty of the nation is being breached in both the case of a drone strike and a full military invasion, both methods breach the sovereignty of the nation affected.

- Another ethical theory to explore would be utilitarianism and how fewer troops on the ground would mean that individuals were less likely to die during the invasion. Less death would therefore mean less loss of utility. This argument can also be made with regards to the greater accuracy of drone strikes and the ability to avoid collateral damage.

**Alternative possibilities to explore (slightly more difficult or less persuasive)**
- Is it morally differentiating to kill an individual using a drone as opposed to using a weapon
- Might drone use psychologically affect pilots, removing from the situation enough that they might be able to carry out actions that would otherwise be uncontainable to soldiers on the ground.

*Conclusion*
- Summarize the main points made on each side of the argument the essay.
- You may wish to come to a decision either way, or it is equally fine to sit somewhere in the middle, so long as this is fairly justified


**Is reducing inequality always a good thing?**

**Introduction**
- Equality is something of a 'buzzword' but, as with many topics in philosophy, unless you actually consider the theory carefully then what might seem like a question you feel you could easily answer is in fact more challenging than you thought. Consider that if two persons each were almost entirely equal except that one of them had a piece of chocolate then if we always prioritise equality we ought to take the chocolate away from this person even though it does no harm to the other individual by the first person having this chocolate. This is the levelling down objection.
- Ultimately, this question comes down to whether equality has any inherent value – it needs to discuss whether there is any value to reducing the happiness of one person to make them equal with another in abstract in order to reach the very highest level

*Arguments for:*
- Equality is something that we strive for in all different parts of life. It is certainly something that we wish to apply to all humans at the point of birth and it might be said that this desire for equality should continue forward.

- There is inherent value to equality between individuals, consider the example of two children who live as perfectly equal siblings until their mother gives one of them a lollipop and not the other one – the average person would see this as unfair and would prefer no lollipops over an imbalance of lollipops between the children

Arguments against
- The levelling down objection shows that a reduction in one persons welfare for no reason other than to make them more equal with a second person is in fact not defensible
- Arguments in support of equality actually misinterpret what it is to accurately discuss the levelling down objection, as there is often a misconception that achieving equality might make one of the parties happier. In fact, this is not the case in the levelling down objection as in order to consider the inherent value of equality in abstract; it is necessary to consider whether it is justified to lower the happiness of one person only for the sake of achieving equality. When one consider arguments for equality with regards to this it becomes clear that the levelling down objection does in fact discredit the idea of equality having inherent value

### *Conclusion*
- A summary of levelling down arguments and objections
- Either a stated conclusion on the value of equality or some degree of uncertainty and an explanation of why you feel that you cannot come to a conclusion

## Is a referendum a good way to decide a major question facing a country?

**Introduction**
- This question is a really good opportunity to bring in examples of current affairs that you think are relevant – this does not mean however that examples are a replacement for analysis and you need to makes sure that you state why examples you choose are directly relevant to the question.
- A common fault with questions like this is just talking about all of the things that you know about referenda rather than actually attempting to focus on what the question is asking

- *Arguments for:*
- Applying a democratic decision making procedure means that the electorate are given an opportunity to engage with issues affecting them in a way that equally values all citizens of voting age
- Referenda create a very energised political atmosphere – seen especially around the Scottish and EU referenda. This motivates individuals to engage in politics, including after the referendum itself and also means that more persons attempt to educate themselves on relevant issues
- If significant decisions need to made and it is in the middle of a government's term and therefore a general election cannot be held on that particular issue, a referendum allows the government to obtain a mandate to carry out action without the need for an election

**Arguments against**
- Referendum campaigning is a prime opportunity for fear mongering and populist arguments rather than the reasoned argument that is more prominent in a parliamentary debate – this would be a great time to bring in examples from recent referenda
- The cost of a referendum is very high and it takes an extremely long time to carry out meaning that quick decisions are very difficult and costly to make through the use of referenda.
- Referenda can be counterproductive if they are not carried out properly – the EU referendum seems to provide a mandate for the government to leave the European Union but since the result there have been a huge number of people who have changed their minds. The 'mandate' is therefore arguably out of date and counterproductive to what the public now may want

*Conclusion*
- Summarize the main points made on each side of the argument the essay.
- In this essay in particular, it might make sense to argue that the value of referenda varies considerably based upon the state of the country, the conduct during the campaigns and the question being asked
- You may wish to come to a decision either way, or it is equally fine to sit somewhere in the middle, so long as this is fairly justified

# SECTION TWO 2017

**Can we learn about intelligence by studying how humans and other animals learn?**

**Introduction**
- This question is asking whether there is value in the study of humans and animals learning and as such, this question is quite challenging for those who do not have any notion of psychological theory
- Despite this, there are general arguments that can be made regarding learning and intelligence but they are unlikely to allow a candidate to access the higher marks without the addition of specialist psychological knowledge
- This question raises two points – the first is whether intelligence can be directly linked to learning and the second as to whether in practice it is possible to learn using studies of learning
- Historically there has been a separation of the study of learning and of intelligence which is an anomaly in scientific psychology
- Increasingly though, there is a claim that there are no clear distinctions which can be made between the cognitive processes that contribute to individual differences in these two areas

**Arguments for**
- Intelligence is expressed relative to the ability of the individual to learn. Smart people tend to learn faster and also learn more than less intelligent people. Intelligence is also clearly manifest in the ability to acquire complicated skills and excel in their performance through practice and progressive improvement.
- The results of recent analysis are consistent with the conclusion that performance on learning tasks and conventional tests of intelligence, e.g. IQ tests, both reflect common factors, for example Spearman's $g$, or the general factor common to all cognitive abilities. For this reason the historical separation between learning and intelligence as psychological concepts ought not to be so rigorously enforced.
- Since learning is dependent upon the existence of systems through which information can be interpreted and translated into skill, systems that we understand as intelligence, the two concepts are fundamentally linked
- By examining differences in the learning processes of intelligent and less intelligent individuals we can then see the difference that intelligence makes to the learning process

**Arguments against:**
- ➤ It is extremely tempting to equate knowledge with intelligence and this needs to be avoided in order to accurately gather information on intelligence through the process of learning. It is accepted that knowledge and intelligence are entirely separate things and that, to an extent, an individual in possession of one is not necessarily in possession of the other
- ➤ In terms of examining animals, it is extremely important not to adopt an anthropocentric standpoint when attempting to evaluate the intelligence of animals through their practices and their ability to learn. It is accepted that animal abilities differ considerably to human abilities and what might indicate a genius intellect in a human might be a skill possessed by all members of a particular animal species. As such there is a certain amount of discretion that needs to be applied when attempting to acquire information about the intelligence of an animal from their learning practices.

*Conclusion*
- ➤ Summarize the main points made on each side of the argument the essay.
- ➤ In this essay in particular, it might make sense to argue that the value of referenda varies considerably based upon the state of the country, the conduct during the campaigns and the question being asked
- ➤ You may wish to come to a decision either way, or it is equally fine to sit somewhere in the middle, so long as this is fairly justified

**END OF PAPER**

› SECTION ONE    2018

# 2018

## Section 1

**Question 1: B**
A is incorrect because the passage suggests technology has made us live longer, not enabled us to choose when we die. C is incorrect because the passage describes a 'new' fear, not a worse fear. D and E are not relevant to the passage and therefore B most accurately expresses the idea that people should be allowed to be euthanised if they wish to.

**Question 2: D**
If we say there are y girls and x boys in the class.
$$3x + \frac{y}{3} = 24$$
$$x + y = 24$$
By solving these equations simultaneously we obtain x=6 and y=18. Then 18-6=12 and so D is the correct answer.

**Question 3: E**
The structure of this argument is as follows: let maize be A and soybeans be B. A is preferred to B yet A is unavailable so the farmer settles for B. E is the only statement which mirrors this because the bank is better than the hotel however the bank was unavailable so they had to use the hotel.

**Question 4: D**
A is incorrect because the statement does not suggest fish farmers should switch, it simply states that carnivorous fish are less environmentally friendly to farm. D is correct because the statement says that salmon can only be farmed by depleting wild fish stocks which means they are worse for the environment than non-carnivorous silver carp.

## Question 5: B
B is correct here because the statement suggests that in order to solve the problem of failing to hire talented researchers and lecturers, the new policy will help to 'improve recruitment' by using independent HR in hiring committees. For this to be effective the HR consultants must be able to spot suitable candidates, if they are unable to do this, they will not improve recruitment procedures and so the argument is flawed.

## Question 6: D
After the spillage, the 400ml is still 4/5 water and 1/5 concentrate. Therefore, 320ml water and 80ml concentrate. Adding 100ml of concentrate means the final amounts are 320ml water and 180ml concentrate.

$$\frac{180}{500} x100 = 36\%$$

## Question 7: C
If we calculate the net loss of professionals by looking at the P column under professionals.
2017: 65-44= 21 out
2016: 67-45 = 22 out
2015: 68-44 = 24 out
2014: 50-55 = 5 in
2013: 51-59= 8 out
Therefore we can see that 2015 had the greatest loss of professionals.

## Question 8: C
This is the correct answer because if we look at the top of the platform clock, the first number must be a 1 and the second must be a 4. In the minutes section the first number must be a 5. Therefore automatically we know the time is 14:5?, whatever the final digit this rounds to 3 o'clock.

## Question 9: E
E is the correct answer because the main conclusion of the argument is that shooting overpopulated species can have very positive conservation effects. However, E suggest that the reintroduction of top predators is more effective and so weakens the statement in the argument.

# SECTION ONE       2018

**Question 10: D**
D is the correct answer because B and E are irrelevant to the passage. C is not necessarily true as the mother seems to be implying good grades may have been achieved despite taking the holiday not as a result of it. A is not related to the content of the argument. D is correct because the mother suggests the law is unnecessary simply from her own experience.

**Question 11: E**
E most accurately expresses the main conclusion of the argument because the argument highlights several flaws in the idea: not only that it punishes single parents but also subverts incentives and punishes those who have not found a partner. Therefore, E is correct as it is the most representative.

**Question 12: D**
From 2011-2012, we can round down 4.2 to 4, $\frac{4}{3}$ is roughly a 33% increase.
From 2012-2013, 6.3-4.2=2.1 so he increase is 50%.
From 2013-2014, 9.45-6.3=3.15 another 50% increase.
From 2014-2015, approximating the figures, 15-9.5=5.5 which is greater than a 50% increase.
From 2015-2016, approximately 22.5-15=7.5 which is roughly a 50% increase.

Therefore, overall, from 2014-2015, the percentage increase is the biggest.

**Question 13: D**
Let us calculate the costs with each wall paper type.
Woodchip- 27 rolls at £25 each, 5 rolls are free, so the total is $22 \times 25 = £550$
Vymura- 18 rolls at £35 each, 10% off is applied, so the total cost is $18 \times 35 \times 0.9 = £567$
Anaglypta- 18 rolls at £40 each, 20% off is applied so the total cost is $18 \times 40 \times 0.8 = £576$
Embossed: 14 rolls bought at £50 each, 25% off is applied, so the total cost is $14 \times 50 \times 0.75 = £525$
Fabric effect: 11 rolls at £100 each, buy one get one free is applied so the total cost is $6 \times 100 = £600$
Therefore the cheapest overall is D, embossed at 525.

## SECTION ONE — 2018

**Question 14: C**
As there are 8 folded sheets on each block, on each half of the block the following page numbers will be shown because each page is double sided:

Block 1 Half 1 - P 1-16
Block 1 Half 2 - P 17-32
Block 2 Half 1 – P 33-48
Block 2 Half 2 – P 49-64
Block 3 Half 1 - P 65-80
Block 3 Half 2 - P 81-96

Therefore, the centre of the block will show pages 48 and 49

**Question 15: C**
C is the correct answer because the statement says that by awarding a Nobel Prize within Economics, this wrongly categorises it as an exact science and further gives examples of why economics does not have the same certainty and predictive power associated with the exact sciences and therefore cannot be treated as one.

**Question 16: B**
B is the correct answer because the argument states that injuries/deaths due to tiredness are preventable by not driving if extremely tired. However if drivers cannot recognise they are dangerously tired, this mechanism will not work to reduce deaths/injuries therefore making the argument invalid.

**Question 17: B**
B is the correct answer here because A may be true but the argument is still valid as people may not use such open spaces, so A is not correct. The article states an increase in the number of children with rickets and so this may be true even if the incidence is rare, therefore C is incorrect. D and E are not relevant to the argument. Therefore B is the correct answer.

**Question 18: B**
If 4 girls have 22 shells each, let us say that the youngest girl has x. The 4 girls each lose 3 shells so now 4 girls have 19 shells each and the youngest has $x + 12$. If they all have the same it must be true that $x + 12 = 19$ and therefore $x = 7$

# SECTION ONE       2018

**Question 19: D**
If we look at training, we can tell that Graham cannot win the prize.
Analysing penalties, David and Colin are also excluded from winning.

John and Mike are the only two left, so by looking at the goals column, Mike scored the most so Mike wins the prize.

**Question 20: B**
Immediately we can see that cube 2 is possible as if the W is folded round, it will resemble cube 2. Therefore C,D and E are incorrect. Cube 3 is also possible as if the A and D are folded down the net will resemble 3. Therefore B is correct.

**Question 21: A**
A is the correct answer because the argument states that phones are supposed to keep teenagers safe but due to cyber bullying and stalking, they have now become unsafe. But phones still make teenagers safer than they were before, in other words, cyber bullying and stalking does not undermine the respects in which mobile phones ensure safety. Therefore A is correct.

**Question 22: B**
B is the correct answer because the structure of the reasoning is as follows: customers are rewarded if they achieve X, you have not achieved X and so do not get the reward. In B the dessert is the reward and the condition of achievement is completion of the main course and so this exactly mirrors the above reasoning.

**Question 23: E**
E is correct because the argument shows the negative impacts of the media on the general public. Magazines are another media form which can have negative impacts on young people and so reflects the principle of the argument.

**Question 24: D**
If we model the time as ab:cd and we want to maximising the number of segments in d, remembering we can only choose between numbers 0-9, 8 has the most segments ( 7). Maximising c, between numbers 0-5 , 0 has the most segments (6). Maximising b, we between numbers 0-9, 8 has the most segments ( 7). Maximising a, between numbers 0-5, 0 has the most segments with 6. Therefore, the time 08:08 is shown which has 26 segments in total.

# SECTION ONE    2018

**Question 25: D**
Round all the valued to the nearest thousand, by adding them up we find that Val and Pat have the two most votes. By adding up each of their scores using long addition (discounting the two scores where they apply to both people). We find that Pat has roughly 600 more votes than Val and therefore Pat is elected president.

**Question 26: E**
From Amy's view point let us label the lines (rows) of piles x,y,z,a. Row x has one pile with 5 boxes; the tallest pile in row y has 6 boxes; the tallest pile in row z has 7; the latest pile in row a has 5.
The far left column from Ben's view must therefore have 6 boxes or less, so E cannot be Ben's view. Therefore E is the correct answer.

**Question 27: D**
The statement says that unemployed people all respond differently to different solution measures. Therefore in order to solve the problem there must be more skilled staff in order to tailor solutions to each individual person. D is the correct answer.

**Question 28: A**
A is the correct answer because the statement makes several points which imply prime ministers should sack ministers to show integrity and to remain In power as well as reflect high moral standards.

**Question 29: A**
A is the correct answer because the librarian pledges to increase the number of computers in the library as a result of student complaints. The librarian need not respond to this in such a way, but they are aiming to cater to the students preferences, therefore A is correct.

**Question 30: C**
Let x equals the number of 9p coupons and if y is the number of 14p coupons, and z the number of 20p coupons. As a result, we have the equation $14y + 9x + 20z = 150$.
We also know that $x > y > z$. We will start by finding combinations of multiples of x and y which are equal to 130.
Starting off with $x = 8\ and\ y = 4$, $14(6) + 9(4) = 120$, this is too small.
Next with $x = 8$ and $y = 2$, $14(8) + 9(2) = 130$ so z must therefore equal 1 giving the combination of 8+2+1=11

**Question 31: E**
'Beginner's guide to history' and 'All about history' are not relevant because we are looking at 1900-1950.
The cheapest book therefore is History of the 20$^{th}$ century and so she withdraws £45.

Now working out 20% off, History of the 20$^{th}$ century is 45x0.8=36, The 'illustrated guide to history' is 50x0.8=40 and 'History for all' is 60x0.8=48, so she buys The illustrated guide to history at £45 leaving £5 left.

**Question 32: E**
The left item has one large diagonal edge and the item on the right has a larger and a smaller diagonal edge. In combination E, we can see both the large diagonal sides for each item, therefore the longest horizontal side on the item to the right is connected to the item on the left, but then the small diagonal side must also be visible. Since it is not visible, combination E is not possible.

**Question 33: C**
C is correct because the new health guidelines suggest 7 portions rather than 5 should be the norm, however a change in target would not result in a change in behaviour as previous evidence has shown. Therefore, current guidance should be maintained.

**Question 34: C**
C is the correct answer because the statement says that the tougher restrictions on eligibility resulted in a fall in the number of recipients. This is an example of confusing correlation with causation and therefore undermines the argument which suggests that tightening these restrictions even more will result in the same effect.

# SECTION ONE  2018

## Question 35: D
D is the correct answer due to the fact that the author states 'More likely this status derives from a small number of his plays, the four great tragedies' rather than the plot development/skill with language.

## Question 36: B
Profit = TR - TC, and revenue = P(price)xQ(quantity sold). Let us find the profit for each item. Always round up for wholesale packs bought leaving some leftover.

|  | Revenue | Wholesale Packs bought | Total Cost | Profit |
| --- | --- | --- | --- | --- |
| Crayon | 180x10=1800 | 180/12=15 | 15x48=720 | 1800 - 720 = 1080 |
| Felt tips | 150x12=1800 | 150/12=12.5=13 | 13x72=936 | 1800 - 936 = 864 |
| Pencils | 200x6=1200 | 200/24=8.3=9 | 9x24=216 | 1200 - 216 = 984 |
| Pens | 150x15=2250 | 150/36=4.1=5 | 5x36=180 | 2250 - 180 = 2070 |
| Rulers | 40x30=1200 | 40/12=3/3=4 | 4x60=240 | 1200 - 240 = 960 |

Pens and Crayons produced the highest profit out of the five items sold

## Question 37: A
It is not necessary to look at the photo development time of the first five people beause they are able insert their money while the photos are being developed. Therefore, the first person is finished in the booth after 3 minutes, the second is finished after 6 minutes and if we continue this we can tell that the $5^{th}$ person is finished after 15 minutes. Therefore, although you have wait after having inserted your money for your photos to be developed, the total waiting time before you can insert your money is still 15 minutes in total.

## Question 38: B
The answer is B because the smallest x value is Hilltop, so the point furthest left is 763613 . Since Longwood is 787634, it is 2 right and 1 up from the square that Hilltop is in. Therefore it has to be W, so B is the correct answer.

## SECTION ONE    2018

**Question 39: A**
The statement implies that problems occur when autopilot systems hand responsibility back to the pilot and this can lead to crashes. Therefore by developing autopilot so that this does not occur, this will reduce the incidence of crashes. Therefore A is the correct answer.

**Question 40: E**
E is the correct aswer as the author wants children to enjoy poetry rather than analyse it and suggests the way to do this is by requiring children to learn and recite poems by heart. But if this does not increase their appreciation for poetry, it will not achieve the author's objective.

**Question 41: C**
C is the correct answer because the statement suggests that unpaid internships mean that less wealthy students are unable to support themselves and then this prevents them from gaining entry into industry. However, if unpaid internships last a few days, students will not have a problem with supporting themselves for this duration.

**Question 42: E**
To reach our total of 10 digits, 4 numbers must have 2 digits and 2 numbers must have a single digit. Let us think of the optimal scenario, the largest number should ideally end with a 0 and the smallest number should ideally end with a large digit. Since the largest number has 2 digits, this leaves 3 other numbers with 2 digits and 2 numbers that have a single digit. Therefore, the 2 single digit numbers should be as large as possible, 8 and then 9. The next 3 numbers which are all 2 digits will have the first digit ascending and the second digit descending. Therefore, we are left with; 8,9,17,26,35,40.

So the smallest range is 40-8=32

**SECTION ONE**  2018

**Question 43: C**
One 8h ticket = £6.20.
Returning increasing quantities of smaller length tickets, one 7h + one 1h ticket = £6
6h + 2h = £5.40
5h + 3h = £5.20
4h + 3h + 1h = £4.90
4x 2h = £4.80
(2x 3h) + 2h = £4.60
Therefore, £4.60 is the cheapest amount for parking for 8 hours, the minimum amount of time valid for a 7.5-hour period. Therefore, £6.20 - £4.60 = £1.60

**Question 44: D**
All 3 faces have 1 triangle shaded and all 5 possibilities only have the middle 3 cubes shaded so there is only one possible orientation for each one. If you look at the two squares on top of each other in each net, and look for the connected edge of those two squares, the connected left vertex of that edge is where all the all the one triangle faces meet and as shown in the picture, 2 triangles meet at this corner. Therefore, we can discount 5 and 2. Also we discount 3 as none of the triangles meet across faces. So we are left with 1 and 4. By visualising 1, we cannot turn this net to match the one in the question and therefore the answer is D.

**Question 45: D**
This is the correct answer as the chain of reasoning in the statement goes as follows. Charities do not attack the root of the problems that market economies create, therefore it is immoral to support them as in doing so this makes the problem deeper. Therefore giving to charity is immoral.

**Question 46: D**
If we label – understanding Pushkin's poetry as A and having a mastery of Russian as B. A only applies conditional on B. B is necessary for A. The second part of the statement states that A is true, therefore B must be true. Therefore D is correct because it follows this structure with A being having a bypass and B being creating pedestrian only high street for shoppers.

# SECTION ONE  2018

**Question 47: A**
A is correct because the main point of the statement is the abuse of power of rail contractors by using inmates as cheap labour and this resulting in an infringement of their rights. The point about safety is more of a sub-point and therefore is not the general principle of the argument.

**Question 48: D**
Let us analyse the first leg of the journey, remembering that speed = distance / time, we can calculate the time as: $\frac{15}{60} \times 60 = 15$ minutes.

Now, if we look at the second leg of the journey, this took a total of 5 hours.

And finally the third leg the duration of which we can calculate: $\frac{20}{40} * 60 = 30\ minutes$, therefore the total time of the journey was 5 hours and 45 minutes. We know that the destination was reached by 12:00 midday, so the journey must have begun at 6:15am.

**Question 49: C**
Firstly, the shelves must be at least 40cm in depth, so we know 30cm wood cannot be used. We should look at the longer lengths to find the cheapest option. Consider a 4m long piece, which can be used to make two 1.8m long pieces. This can make 4 of the selves if we buy 2 x 4m at 45cm because this is the cheapest option. If we buy a 2m piece at 45cm, this account for the shelf left over.

$$£9.30 + £9.30 + £4.95 = £23.55$$

**Question 50: C**
Combination A will be symmetrical because the pattern changes such that the bottom line becomes horizontal and the others will cross in the middle making them symmetrical.
Combination B, the bottom two lines on the left will create a symmetrical cross at the bottom. The top line on the left will be horizontal and the middle lines will form another symmetrical cross making B. If we think about combination C, we only need to consider the top line, this line will be diagonal linking the top left and bottom right and there is no line to symmetrically match this. Therefore, C is not symmetrical.

## END OF SECTION

# Section 2

**Is humanism a religion?**

**This essay should contain:**

- Humanism is a moral ideology which posits positive attitudes centred around our common humanity
- What is a religion? How can we define religion? The answer must define religion with certain criteria and test whether humanism adheres to such criteria
- Consider both sides of the argument and weigh them up to reach a final conclusion
- Should be less superficial than a simple science vs religion argument, it should delve deeper into the psychology behind religion and human belief.

**Suggested points:**

- Humanism suggests a moral code to live by, a feature of religion is the creation of a code of conduct to live by, normally which involves revering/respecting a deity
- Even through humanists do not agree with religion, it is a system of beliefs itself and so can be regarded as a form of a religion
- Religions have many supernatural elements but humanism is based on the real world
- Humanism only makes scientific claims and does not require belief without concrete proof
- Student could may suggest texts like JSM's 'On Liberty' and Richard Dawkins/ Bertrand Russell literature as a 'religious text' of humanism

**Should journalists only be allowed to follow their profession if licensed to do so, like medical doctors?**

**This essay should contain:**

- Meaning of lisencing: under what conditions?
- The answer should explore what licensing entails and comparisons between the implications of licencing for both medicine and journalism
- Introduction which defines key terms and identifies line of thought for their argument in a concise way
- Conclusion which summarises the argument, opening up avenues for new research

**Suggested Points:**

- Journalists need to be able to retrieve information from reliable news sources and present stories with strong evidence and transparency
- Fake news has become a growing problem in society, which can result in the mis-education of the public. Licensing of journalists may reduce the potential for this to occur
- Licensing could allow government capture to occur within the media industry – as who would they be licenced by? It must be an independent body, but who would fund them?
- Licensing may deter potential journalists if they cannot afford to obtain a licence
- It seems unnecessary to provide journalists with licences because the responsibility associated with the occupation is not as important as that of a doctor

**Should a government impose a legal maximum ratio between the highest and lowest pay of individuals in companies?**

**This essay should include:**

- ➢ Technical economic analysis and chains of thought, as this question is geared towards economists
- ➢ Introduction should outline the argument and introduce the candidates reasoning
- ➢ Conclusion should accurately summarise the argument
- ➢ Essay should consider whether this is a viable economic policy and what the potential effects could be
- ➢ What would be the motivation for the introduction of this policy

**Suggested points:**

- ➢ A maximum ratio could result in lower productivity within firms as shareholders are unable to incentivise greater performance from the highest-earners within the firm
- ➢ However, this may not necessarily the case, as evidence has shown that firms which have an employee-centric business model may grow faster than others
- ➢ Inequality within a firm may lead to new recruits becoming disillusioned when comparing their salaries to those of the CEO/managers
- ➢ Firms may move to a different country to avoid this policy and it will therefore result in 'brain drain' and hence, lower GDP growth
- ➢ The ethical motivation for this policy should be considered- is this fair?

**SECTION ONE** 2018

**Do non-human animals have opinions? Do they have beliefs?**

**This essay should include:**

- An introduction which outlines the key points within the candidates essay
- Confirms the definition of words in the statements: i.e opinions could be interpreted as expressing like/dislike towards certain things
- The argument should follow on the basis of this definition (i.e the argument depends on the definition the candidate chooses), the candidate may explore multiple definitions to show breadth
- The conclusion of the essay should state the candidate's final viewpoint on the question and perhaps suggest opportunities for greater research on this topic

**Suggested Points:**

- Observations amongst elephants and certain animals have shown a trend towards family units and groups rather than acting alone. This could be interpreted as an opinion
- Are animals conscious? Perhaps pronounce religious viewpoints on this
- Animals have been shown to express empathy in the wild – documented account of a humpback sweeping a seal onto its back away from killer whales
- Animals can communicate and this, in itself, is a form of expression
- Given that we have evolved from animals (apes), many species have an extremely similar genetic makeup to humans and so, they are
- Can animals tell the difference between right and wrong? Does this indicate that they have moral codes of conduct

**END OF PAPER**

SECTION ONE  2019

# 2019

## Section 1

**Question 1: E**
The passage is discussing the increasing property prices in London and its effect on the city's prosperity. **E** sums this up the best as it highlights **both** the cause (unparalleled increase in property prices) and the effect (London's damaged prosperity) the passage is discussing. Thus **E** is the main conclusion.

**Question 2: C**
The amount of tax that **Tim** pays is 30% of $10000 = 0.3 x 10000 = $3000.

Going through each option we notice **David's** tax payment is 6% of $50000 = 0.06 x 50000 = $3000. This is the same as Tim. Answer is **C**

Note: that 6% is 5x less than 30%, **and** 50000 is 5x more than 10000 so the 2 cancel each other out.

**Question 3: E**
Answer is not **A** as passage says self-discipline is **one** quality not the **only** requirement.
**B** is irrelevant because the article discusses developing self-discipline skills to apply to studies not for studies.
**C** is not the flaw as the article is not implying the need for elite level coaching to develop self-discipline.
**D** is completely irrelevant to the argument
**E** is the flaw in the article as it assumes that self-discipline is the effect of training rather than a something naturally possessed by athletes. Thus advocating for young people to undertake training programs may be pointless

**SECTION ONE** **2019**

**Question 4: B**
Not **D** as passage doesn't mention about people dying from drink/ drugs
We can't conclude **E** as it could be a generational factor than an age factor.
Not **C** as no link discussed between drink and criminal activity in the passage.
Not **A** as while they have fewer health problems due to drink/ drugs young people could have more net health problems due to other dominating factors such as stress or allergies.

**B** is correct as the passage discusses social problems are becoming more prevalent for the middle age, 39% of heroin addicts are now 40 or older compared to 19% in 2006.

**Question 5: A**
This passage is arguing that because teachers only teach for exams rather than preparing students for life then education is **less valuable**. The passage is **assuming** that teaching for the sole purpose of exams is, therefore, less valuable than teaching to prepare a student for life. **A** highlights this assumption.

**Question 6: D**

If we say the width of the flag is 4 and the height of the flag is 2.

Area of the red rectangles is 2 lots of 2 x 1 = 4
Area of the white square is 2 x 2 = 4.

So area of leaf is ¼ of white square which is ¼ x 4 = 1 so total area of red = 4+1 =5 and area of white is 4 -1 (area of maple leaf) = 3

So ratio of red to white in the flag is 5:3. Therefore Answer is **D**

Note total area of flag is (1+2+1) x 2 = 8 and Area of red + Area of White = 5+3=8

# SECTION ONE    2019

**Question 7: E**

Numerically, the pair of examiners who disagreed most on the marks awarded will be the pair that has the largest difference between the two marks awarded.

The largest difference for a student is 40 (= 90 − 50) and that is for **Hilary Gordon**. The pair of teachers that marked her paper was **Mrs De Vere and Mr Robson.**

Therefore, the answer is **E**

**Question 8: E**
Let S = Standard and P = Premium, we know price of S < price of P
On Thursday: we have 200 S + 150 P
On Friday:   we have 150 S + 230 P,
Change from Thursday to Friday is − 50S and + 80P as S<P and 50 < 80 we know that the total income for Friday will be strictly greater than the total income for Friday.

So Thursday can never be the day with the highest sales income so **not A, C, D**

On Friday:    150 S + 230 P
On Saturday:  300 S + 120 P
Change from Friday to Saturday -150S and + 110P now we know S < P but 150 > 110 so we don't know if -150 S + 110 P > 0 (i.e. Saturday sales are higher than Friday) it will be true is 110P > 150 S so P > (11/15) S but as we don't know the prices of S and P we only know P > S but not by how much we can conclude which one is higher. But we know it's definitely not Thursday. Therefore, the answer is **E**

**Question 9: B**
The passage discusses the perceived harm of texting on correct grammar and the outcomes of having poor grammar. Since mobile phones have been banned test scores have gone up. The Author **assumes this is causal** when it may be spurious or both are the outcome of something else. I.e. a stricter teacher who is more determined for students to do better on their tests. The flaw is **B**

## Question 10: E

The article discusses that if changes are known to be made by a woman then the acceptance drops to 62% from 72% when the woman's gender is unknown to others. This suggests that there is a negative bias in the acceptance rate of code changes when people know it is being made by a woman **specifically**.

If **E** were true then it is not the case that knowing it is a **woman** has changed the code but leads to a drop in the acceptance due to the revelation of **any gender** so there is no specific bias against women's acceptance rate because they are a woman.

## Question 11: D

Passage is discussing that while some kids don't enjoy the competitive nature of sport, being active is important for general health at all stages of life. Many people stop sport when they leave because they didn't enjoy the competitive nature. So schools should offer non-competitive alternatives to traditional games lessons to make it more likely they continue to be active once they leave. Encourages being active rather than engaging in competition. The answer is **D**

## Question 12: C

If we look at a 4 second snapshot

| Second | First Lamp,1 off 1 on | Second Lamp, 2 off 2 on |
|---|---|---|
| 1 | ON | ON |
| 2 | OFF | ON |
| 3 | ON | OFF |
| 4 | OFF | OFF |

We are looking for when they are both off. In a 4 second period they are both off for the 4$^{th}$ Second. Thus 1 out of every 4 seconds they are both off. There are 15 Four-Second intervals in one minute. So in one minute they are **both off** for 15 seconds. Therefore, answer is **C**.

Embossed: 14 rolls bought at £50 each, 25% off is applied, so the total cost is $14 \times 50 \times 0.75 = £525$
Fabric effect: 11 rolls at £100 each, buy one get one free is applied so the total cost is $6 \times 100 = £600$
Therefore the cheapest overall is D, embossed at 525.

## SECTION ONE    2019

**Question 13: C**
If we assume that average monthly wage in each country is 100. Then we just need to check which fraction is smallest. Where 'with cost as a %' is the **numerator** and **length** is the denominator.

14/56>8/69 so **not A**
8/69 > 6/75 so **not B**
6/75 < 6/72 so **not D**

We must compare 6/75 and 5/46

6/75 = 2/25

2/25 *(2.5/2.5) = 5/62.5 < 5/46 so not **E**

Therefore 6/75 is the smallest so the **answer is C**.

**Question 14: C**
By looking at first graph. Car 2 is newer and more expensive. By looking at the second graph Car 2 is faster and bigger.

So which statement is **not** correct?
The cheaper car is car 1 but the bigger car is car 2 so clearly **B is not correct.**

**Question 15: B**
Passage discusses how a recent flood management scheme has helped save some properties. Then it expands on the specifics of the ways in which floods are managed. Then discusses **other benefits** of the land development such as improved water quality, preventing erosion etc.
Therefore, we can draw **B** as a conclusion.

**Question 16: A**

The article suggests that the government must abandon plans to raise fuel tax. This is because meeting carbon emissions targets will have large negative impact on the economy. The **assumption** the article is making to suggest abandoning the plans is that the economy is more important than the need to control carbon emissions. Thus the answer is **A**

# SECTION ONE  2019

**Question 17: E**

The article is suggesting that by virtue of being married, couples have improved health, wealth, and are happier. However, if **E** is true then the causality runs the other way. Wealthier couples are more likely to marry rather than the state of being married leads you to be wealthier, the people are wealthier from the offset.

Hence **E** if true most weakens the reasoning in the passage.

**Question 18: D**

Alec and Colin will cross the start line every other lap which will be every 6 minutes.

We need to find when Colin and Barry cross the start line as every time Alec does Colin will.

LCM of 6 and 5 is 30.

Thus in 30 minutes they will all cross the start line at the same time. In those 3 minutes Alec would have completed 10 laps and Barry will have completed 6 laps.

The **difference is 4**.

The **answer is B.**

**Question 19: E**

It is most efficient to work biggest to smallest as we are looking for the biggest increase. So let's first see if we can find a percentage increase of 200%. A percentage increase of 200% means the end value is 3 times as more as the starting value.

So a 200% increase on 1 is 3.

**Grandia** has gone from 20 to 60 which is a 200% increase.

**Therefore, the answer is E**

## Question 20: C
We could see **A** if we looked from the front.
We could see **B** if we were looking at it from the left side.
We could see **C** if we looked at in from underneath.
We could see **D** if we looked at it from the back
If we looked at the shape from above we could see something similar to **E** but there would be no solid line all the way along the top. Along the top of the 'notch' (as is in **C**) there would be no solid line on the part. So answer is **E**.

## Question 21: A
This article discusses the success of conservation tourism in Africa (saving endangered species). Therefore, because it has been a success in Africa it **will** be a success in the Arctic. The flaw is that it makes a general rule from one example, therefore the answer is **A**.

## Question 22: E
The logic is as follows.
If A then B. We observe B so A must have happened,

- **A.** If A then B. Not B so not A
- **B.** If A then B. A so B
- **C.** If A then B. Not B so not A
- **D.** If A then B. Not A so not B
- **E.** If A then B. We observe B so A must have happened.

Therefore the answer is **E**

## Question 23: B
The principle of this argument is: People who advocate for X should not complain if X happens and in doing so there is some negative impact to them. 'Not in my backyard'.

**B** follows the same reasoning as the passage.

## SECTION ONE    2019

**Question 24: C**
If we draw out each digit from the seven segment display then we and see which segments appear the most and which the least. The segment that appears the most is the **bottom right**. It appears **8 times**. So we need to check if the minimum is less than 5 (therefore C) or exactly 5 (therefore D). The bottom left one appears **3 times**. So the answer is **C**.


**Question 25: D**

We must stay Saturday and pick 3 other nights round those days.
For the top 3 the cheapest nights are **Fri, Sat, Sun, Mon.**
For the bottom 2 options the cheapest nights are **Thurs, Fri, Sat, Sun,**

We must check each option applying the deal correctly. The cheapest option is Palace which has 50% off so costs **€140**. Answer is **D**

**Question 26: D**
Key observation in this question is that if a dice rolls 2 times in the same direction the face that will be on top is the face opposite the starting face. So **triangle is the starting face**. Rolls twice to the right. The face opposite will now be on top (**rectangle**). If it then rolls twice down then **triangle** will be back on top. Then rolls twice the right to its final position. As before **rectangle** will be facing up. Therefore, the answer is **D**.

**Question 27: B**
Article argues that climate has been through a recent phase of stability and during this phase was also the rise of human civilisation. This phase has been an exception rather than the norm. The Earth and nature has always found a way of adapting to changing climate. Therefore, acting against climate change is more important for the survival of humans than it is the environment. **B** captures this main conclusion.

**Question 28: B**

This passage discusses that while there are benefits to some as a result of seeing a chiropractor – as is the intention. However, many suffer negative results ranging in severity including death. Therefore, we could conclude that Chiropractic therapy may do more harm than good. Answer is **B**

# SECTION ONE 2019

**Question 29: A**

This article discusses how **due to high costs** face by drivers for attending speed awareness courses could mean police profiteer of this initiative for their own gain. However, **if** the police are profiteering from the initiative then the costs of running re-education courses is less than the combined contributions of participants which is **assumed** in the article. Answer is **A**

**Question 30: E**

Exploring each option.
- A. If we play first and choose 1 and 8 we would have a score of 7. Playing second would win as they would choose 9 and 7 to achieve a score of 8
- B. Playing first choosing 3 and 9 then we would have a score of 6. Choosing 6 and 7 going second would give winning score of 6.5
- C. Playing first and choosing 6 and 7 for a score of 1. Choosing any pair going second would be a win
- D. Playing first and choosing 7 and 9 for a score of 2. Choosing any pair going second would win.
- E. **Going Second** therefore must be a winning strategy.

**Question 31: B**
There are 10 days which are wither a Saturday, Sunday or Bank Holiday. Therefore there are 31-10 = 21 normal weekdays.

S/S/BH = 10 days

Takes 50 minutes for a crossing then 20 min turnaround

9:30am – 4:10pm is 6 hours and 40 mins which is 400 minutes,
Each crossing and turnaround is 70 mins. The final trip will be 50 mins with no need for turnaround so there will be 70*5 + 50*1 = 400 which is 6 crossings.
50 – 20 – 50 – 20 – 50 – 20 – 50 – 20 – 50 – 20 – 50.
So Total crossing in the month on Sat/ Sun/ Bank Holiday = 6*10 = **60**

Normal Weekdays = 21 days

Takes 50 mins for a crossing then 10 min turnaround.

10:30 am – 2:20pm is 3 hours and 50 mins which is 230 mins,

Each crossing and turnaround is 60 mins. The final trip will be 50 mins with no need for turnaround so there will be 60*3 + 50 *1 = 230 which is 4 crossings,

50 – 10 – 50 – 10 – 50 – 10 – 50.

So total crossing for normal weekdays is 21 * 4 = **84**

Total crossings for month of May = 60 + 84 = **144**. Answer is **B**

## Question 32: C

| Andrew | James | Roger |
|---|---|---|
| -3 | +3 | |
| | -2 | +2 |
| -5 | | +5 |
| | +1 | -1 |
| Total = -8 | Total = +2 | Total = +6 |

Total number of votes = 20

- A. Andrew 10 → 2, James 5 → 7, Roger 5 → 11  OK
- B. Andrew 12 → 4, James 7 → 9, Roger 1 → 7  OK
- C. Andrew 9 → 1, James 13 → 14???
     Should be 13 → 15 but clearly below 15 mark.

So answer is **C**.

## Question 33: A
Article argues that if the traders believe the tax loop holes are 'immoral' they shouldn't exploit for themselves. However, if **A** is true then they are doing it not for personal gain – like large corporations – but to try and encourage the government to change the law and close the loopholes. Thus if **A** is true then it weakens the argument in the passage.

## Question 34: A
Passage argues that high use of mobiles phones **causes** to 'cognitive failures' such as short attention span. However, it could be the case that those with short attention spans possessed those traits **prior** to their phone habits. Furthermore it the causality may be in reverse that those with those traits are more likely to be high users of their mobile phones. Answer is **A**.

# SECTION ONE 2019

**Question 35: D**
Article challenges the view that being a hypocrite is a 'bad thing'. Instead it argues that by being a hypocrite it may show that a person has changed for the better. Their present morals do not align with their previous misgivings. Thus hypocrisy can be considered a virtue. To conclude, it is possible to think of hypocrisy as a good thing. Answer is **D**

**Question 36: B**

We need 4 rectangular pieces of cloth that measure 15cm x 38cm.
Which is a 15cm x 152cm

The prices of the pieces of cloth per cm width is.

| Width | Price **per metre** |
|---|---|
| 100cm | £2.00 = 20p/10cm |
| 120cm | £2.20 = 22p /10cm |
| 160cm | £2.80 = 28p/10cm |

As the material is sold in exact 10cm lengths we want to 20cm of cloth with width 160cm to make sure we have sufficient cloth at the cheapest price. So price will be 2 x (£2.80)/10 = £0.56. Answer is **B**

**Question 37: C**
With this question we want to cross out an option if it doesn't match to the requirements.

These are
- 1600mm x 700mm
- Side Grips – YES
- Pre-drilled taps – NO

The cheapest steel bath that meets **all** 3 criteria is the **Europa** at £126
The cheapest acrylic bath that meets **all** 3 criteria is the **Carola** at £130

Therefore the price difference between the two is **£4**

Answer is **C**

**Question 38:**

Using one piece of information at a time we can rule out certain answers.

Piece of information number 1 (theme park + steam train excursion) must be > ½

For **A** 35 + 55 = 90 which **is not** greater than half of 180. So not **A**

For **B** 24 + 32 = 56 which **is not** greater than half of 120. So not **B**

C, D, E are OK


Piece of information number 2 (farm is greater than museum)

For **D** there are equal at 15 each. So not **D**

C, E are OK

Piece of information number 3 (Theme park + museum + farm) must be > ½

For **C** 65 + 15 + 20 = 100 which is = ½ of 200 so **not C**

E is OK, **Answer is E**

## SECTION ONE 2019

**Question 39: C**
Passage highlights the necessity of teaching proper punctuation, as was the convetion years ago. The passage also criticises the current use of commas, that it lacks 'rigour'. Thus we could conclude **C** as the lazy use of commas implies that we no longer possess 'this kind of attention to punctuation'. Answer is **C**

**Question 40: B**
Article attacks the acceptance of long hours and poor pay as well as the negative effects that accompany them – anxiety and stress. It concludes that 'political slogans have clearly been effective'. They **assume** these have to ability to affect everyday life rather than it stemming from another factor. Answer is **B**

**Question 41: E**
Passage discusses how police, like the public, hold false beliefs about crime and punishment. This is an example of the 'scientist – practitioner gap' and the advocate more training as this has been show to permanently reduce the gap. If **E** is true, then **proper training** will mean police officers will have more evidence based views as they would have received the training the article suggests. Thus **E** strengthens the argument for additional training

**Question 42: E**
Woof surface area of a shed

Front panel = 2.5 x 1
Back panel = 2 x 1
2 x (Side panel) = 2 x ((2+2.5) x ½ x1) = 4.5

Total surface area of the wood = 2.5 + 2 +4.5 = 9

Number of coats = 3
Number of sheds = 2

Number of metre squared to be painted = 9x3x2 = 54

Total time = 54 x 4 (time to pain 1 square metre) + 20 (tea break)

$\qquad$ = 216 + 20 = 236 = 3 hours 56 mins.

Answer is **E**

# SECTION ONE  2019

**Question 43: A**
4 teachers & 20 students = 24 people
8 are 15 years old & 12 + 4 are **16 or older**

So 8 children and 16 adults.

Family ticket is cheaper than 2 adults and 2 children.

Group ticket (8 people at least 2 kids) is cheaper than 2 family ticket.

So we can buy 3 group tickets

Group ticket 1: 6 adults and 2 children
Group ticket 2: 6 adults and 2 children
Group ticket 3: 4 adults and 4 children

Price = 3 x $175
Price = $525
Answer is **A**

**Question 44: C**
- A. Will as the shape will not be a hexagonal prism
- B. Will not, if we rotate the T to the 1 square. The resulting join between the T and 1 will mean the T is at an angle rather than perfectly horizontal to the 1.
- C. Answer is **C** by process of elimination.
- D. Will not be D as with **B** the T will be horizontal to the 4 in the actual configuration. But in D the T will be at an angle that is not horizontal shown by rotating T one position so it is touching the 4 square.
- E. Will not be E as the top of the 5 should be touching the T rather than the right hand side on the number '5'.

Answer is **C**

### SECTION ONE  2019

**Question 45: D**
Passage argues that Fukushima residents should be allowed to move back even though there is radiation if they so desire. The use the example of Chernobyl evacuees having a lower life expectancy due to the psychological pain of leaving their homes rather than from cancer arising from radiation exposure. Answer is **D**

**Question 46: E**

The reasoning is as follows

A $\Leftrightarrow$ B, B $\Leftrightarrow$ C which implies A $\Leftrightarrow$ C not C therefore not A. ( $\Leftrightarrow$ means 'if and only if')

- **A.** A $\Leftrightarrow$ B, B $\Leftrightarrow$ C which implies A $\Leftrightarrow$ C. C so A (So not **A**)
- **B.** A or B. not B so A. (So not **B**)
- **C.** A $\Leftrightarrow$ B, B $\Leftrightarrow$ C which implies A $\Leftrightarrow$ C. C so A (so not **C**)
- **D.** A implies B. Not B so not A (so not **D**)
- **E.** C $\Leftrightarrow$ B, not A $\Leftrightarrow$ not B so A$\Leftrightarrow$ B. Which implies not A $\Leftrightarrow$ not C. Not A as not C **(the ordering is different but the $\Leftrightarrow$ is 'reflexive', therefore a different ordering is not important)** Answer is **E**

**Question 47: E**

The principle of this argument is that if people should be able to use the money that belongs to them for their own advantageous reasons, such as private medicine. Therefore, **E** matches the principle of this argument as by virtue of being the rich person's child the money belongs to them – so it should not be taken by the state – and they can spend it to make a substantial difference in their lives.

Answer is **E**

## SECTION ONE — 2019

**Question 48: D**
8:00 am Monday morning 50g flour and 50g of Water.
**Note 24g of water evaporates every day (1g per hour)**
8:00 am Tueday. 50g flour + 26g Water = 76g so 38g of water and 38g of flour added = 88g F + 64g W

8:00 am Wednesday. 88g F + 40g W = 128g. So 64g of water and 64g of flour added = 152g F + 104g W

8:00 am Thursday. 152g F + 80g W = 232g. So 116g W and 116g F added which = 264g F + 196g W

8:00 Friday 264g F + 172g W = 436. So 218g F & 218g W. Clearly this will be over 550g of total mixture. Thus we have the necessary amount on Friday.

Answer is **D**

**Question 49: C**
Minimum amount of almonds = 20%.

20% of 250g = 50g so 200g left over.

Largest amount of peanut in 250g is 40% = 100g.

If 100g of peanuts and there is 200g left of the mixture after eating all the almonds (50g). Then the largest peanut content will be 100/200 = ½ which is 50%.

Answer is **C**

**Question 50: D**
- A. If A were correct then VI should be where I is on the view. As that corner is the corner of V, III, VI. Not A
- B. The numeral directly above IV is I not II as appears in B. Not B
- C. The numeral either directly above III or directly below III is either VI or I not II. So not C
- D. Answer is D by process of elimination
- E. While III is next to V the orientation displayed in E is incorrect. Relative to the way V is positioned then III should be horizontal not vertical as shown in E. So not E

Answer is **D**.

## END OF SECTION

# Section 2

Should children strike to demand action on a major issue such as climate change?

**This essay should contain:**

- A discussion on the mental capacity of a child to understand why they are striking.
- If they do possess the mental capacity what evidence may there be?
- Even if they don't should children be allowed to strike? What other reasons might there be for children to do it, are these specific to a certain cause such as climate change?

**Suggested points:**

- Do adults really strike for a cause or strike for the purpose of striking? Are they themselves truly aware of the cause? Thus if some adults aren't aware but still strike does it matter if some children aren't aware either?
- Even if they aren't aware of the spirit of striking does it send a powerful message in the case of climate change as they will be the ones to suffer the most?
  - Is it therefore more appropriate for a child to strike for climate change rather than higher pay for academic staff?

# SECTION TWO 2018

Assume that automated face recognition is completely reliable. What restrictions, if any, should there be on its use?

**This essay should contain:**

- What could be implemented due to automated face recognition technology
    - Monitoring its population
        - For Good
        - For Bad
- What is the trade-off between the two?
    - If there is a poor trade-off what restrictions could there be on its use to make the trade-off fair?

**Suggested Points:**

- Is having any restrictions important? If people having nothing to hide then what is the issue?
- Social scoring systems with arbitrary rules
    - You accidently drop a piece of litter your face is recognised and it has a negative impact on your entire life

# SECTION TWO — 2018

**Could somebody choose to change their race?**

**This essay should include:**

- A discussion of the parallels to 'choosing' to change gender.
- Is race the same as gender can people feel as if they are born into a different race
    - If not why not?
- What do we mean by choose? To decide to physically appear as another race or identify with the possible plight of another race?

**Suggested points:**

- Do people actually 'choose' to 'change' race or are they born that way but in the wrong body?
- Will people never understand what it is like to be a different race and therefore can somebody really choose to change their race? And so even if you choose to change your race, are you actually any different?
- Is it important to understand what it's like to be another race or is your own self-expression more important and thus someone can choose to change their race?

## SECTION TWO — 2018

Should the main objective of a business to make money?

**This essay should include**

- Define other possible main objectives as opposed to making money
    - Protection of the environment
    - Helping disadvantaged people
- An argument for why this should be the main objective rather than to make money?
- A discussion that any other main objective and in fact most things in life require money and thus to having the main objective as making money ensures a business' survival and thus allows them to pursue other objectives.

**Suggested Points:**

- How is making money perceived?
- How are the other options perceived?
    - Should they be perceived in that way?
- Are all of these 'main objectives' pointless? Is the only objective of a business is to provide a good or service that if intends to? Making money is a by-product of this and the business can then choose to do what it wishes with the money.

## END OF PAPER

## Afterword

Remember that the route to a high score is your approach and practice. Don't fall into the trap saying that *"you can't prepare for the TSA"*– this couldn't be further from the truth. With knowledge of the test, time-saving techniques and plenty of practice you can dramatically boost your score.

Work hard, never give up and do yourself justice.

Good luck!

## About Us

*Infinity Books* is the publishing division of *Infinity Education Ltd*. We currently publish over 85 titles across a range of subject areas – covering specialised admissions tests, examination techniques, personal statement guides, plus everything else you need to improve your chances of getting on to competitive courses such as medicine and law, as well as into universities such as Oxford and Cambridge.

Outside of publishing we also operate a highly successful tuition division, called UniAdmissions. This company was founded in 2013 by Dr Rohan Agarwal and Dr David Salt, both Cambridge Medical graduates with several years of tutoring experience. Since then, every year, hundreds of applicants and schools work with us on our programmes. Through the programmes we offer, we deliver expert tuition, exclusive course places, online courses, best-selling textbooks and much more.

With a team of over 1,000 Oxbridge tutors and a proven track record, UniAdmissions have quickly become the UK's number one admissions company.

Visit and engage with us at:
Website (Infinity Books): www.infinitybooks.co.uk
Website (UniAdmissions): www.uniadmissions.co.uk
Facebook: www.facebook.com/uniadmissionsuk
Twitter: @infinitybooks7